Focus on Insect Rearing Methodology to Promote Scientific Research and Mass Production

Focus on Insect Rearing Methodology to Promote Scientific Research and Mass Production

Editors

Man P. Huynh
Kent S. Shelby
Thomas A. Coudron

MDPI • Basel • Beijing • Wuhan • Barcelona • Belgrade • Manchester • Tokyo • Cluj • Tianjin

Editors

Man P. Huynh
University of Missouri
USA

Kent S. Shelby
USDA Agricultural Research
Service
USA

Thomas A. Coudron
USDA-ARS Biological Control of
Insects Research Laboratory
USA

Editorial Office
MDPI
St. Alban-Anlage 66
4052 Basel, Switzerland

This is a reprint of articles from the Special Issue published online in the open access journal *Insects* (ISSN 2075-4450) (available at: https://www.mdpi.com/journal/insects/special_issues/rearing_research_production).

For citation purposes, cite each article independently as indicated on the article page online and as indicated below:

LastName, A.A.; LastName, B.B.; LastName, C.C. Article Title. *Journal Name* **Year**, *Volume Number*, Page Range.

ISBN 978-3-0365-2492-4 (Hbk)
ISBN 978-3-0365-2493-1 (PDF)

Contents

About the Editors

Man P. Huynh is a Research Associate at the University of Missouri. His research has focused on insect nutrition and toxicology and its utilization in insect production and management via the application of multi-omics analyses and mathematical modeling. This results in the development of artificial diet formulations specialized for western and northern corn rootworm larvae that are currently utilized in research and commercial applications. His current research aims to develop a high-throughput system for mass production of corn rootworms toward accelerating discovery efforts related to novel insecticidal compounds and their related products.

Kent S. Shelby is a Research Entomologist stationed at the USDA Agricultural Research Service Biological Control of Insects Research Laboratory in Columbia Missouri. His research has focused on improvements to the effectiveness of biological control agents such as parasitic Hymenoptera, beneficial Heteroptera, nematodes, and viruses. His current research is a multidisciplinary effort applying multiomic approaches to studying the nutrition, toxicology, and immunity of maize pests.

Thomas A. Coudron is a Research Scientist (retired collaborator) with the USDA Agricultural Research Service. He has been involved in insect biochemistry and physiology for forty years. His early research focused on ways to improve the effectiveness of pathogens, parasitoids, and predators via exploiting the physiological interactions between beneficial organisms and their hosts or prey. This resulted in the discovery of new enzyme complexes and unique nonlethal and nonparalyzing venom components. His recent research has focused on insect nutrition and the application of molecular technology to advance the formulation of artificial insect diets. This has led to diet recipes that are now utilized in research and commercial applications.

Editorial

Recent Advances in Insect Rearing Methodology to Promote Scientific Research and Mass Production

Man P. Huynh [1,2,*], Kent S. Shelby [3,*] and Thomas A. Coudron [3,*]

[1] Division of Plant Science & Technology, University of Missouri, Columbia, MO 65211, USA
[2] Department of Plant Protection, Can Tho University, Can Tho 900000, Vietnam
[3] Biological Control of Insects Research Laboratory, USDA-Agricultural Research Service, Columbia, MO 65203, USA
[*] Correspondence: manhuynh@missouri.edu (M.P.H.); kent.shelby@usda.gov (K.S.S.); coudront@missouri.edu (T.A.C.)

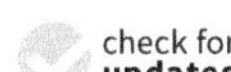

check for
updates

Citation: Huynh, M.P.; Shelby, K.S.; Coudron, T.A. Recent Advances in Insect Rearing Methodology to Promote Scientific Research and Mass Production. *Insects* **2021**, *12*, 961. https://doi.org/10.3390/insects12110961

Received: 15 October 2021
Accepted: 19 October 2021
Published: 22 October 2021

Publisher's Note: MDPI stays neutral with regard to jurisdictional claims in published maps and institutional affiliations.

The benefits obtained from our ability to produce insects have encompassed a wide array of applications, from the early stages of examining different species, to the present day of mass production for multiple purposes. Perhaps the most prominent application to date is insect management for production agriculture. Much of the considerable knowledge humans have of insect biology has been made possible by first bringing insects from the field into the laboratory, and maintaining sufficiently large colony sizes on bespoke, empirically developed diets. The ability to conduct experiments without the seasonal limitations that normally bound insect life history, to control abiotic and biotic treatments, is heavily dependent upon a nutritionally complete, inexpensive and easily produced or procured diet acceptable to the insects. Nominally, healthy insects produced in this way are more likely to respond accurately in various bioassays, to better perform the desired ecosystem services upon release, or to act as nutritional inputs for other animals. Ecosystem services provided by mass reared insects include pollination, release of predators or parasitoids for biological control and releases of sterile insects to suppress pest populations. More recently, the usage of mass-produced insects for nutrient sources such as for human consumption, animal feed, pet food, stock chemicals, or valorization of waste streams has the potential to surpass all previous applications. Of course, food sources are not the minimum requirement for successful mass rearing. Attention must be paid to innovate and optimize many other inputs including reduced labor, close observation of losses at each life stage, mating and oviposition, appropriately shaped containers, lighting, scheduling and sanitation.

The initial effort of the mass production of insects probably has its origin in scientific inquiry. The production of healthy, vigorous experimental subjects for basic and applied research purposes was an early impetus for the development of insectary diets and rearing procedures. That desire to study insects remains and has been greatly assisted by the ability to produce the specimens under natural and controlled conditions. Typical of research efforts is the progression of developing, refining and adapting supportive procedures that enable advancements. Insect rearing is a prime example.

After collection, the need to rear becomes the dominant challenge in all efforts to produce insects and that requires an understanding of environmental and nutritional needs. Along with an increased knowledge of insects came technological advances that resulted in improved methodologies and commercially available supplies. Courses, symposia and conferences in insect rearing were held and as the knowledge-based methodologies and supportive materials expanded, so did the applications of rearing insects. An industry of insect production and sales emerged and continues today, as does the need for quality assurance associated with rearing, research and production.

Numerous insects have been reared and many continue to be reared today. Applications include discovery of insect pheromones, repellents and new compounds with insecticidal properties. Some are used for monitoring purposes, including detection of insect resistance. At the forefront of mass rearing was the emerging field of the sterile insect technique (SIT) for area-wide management of pest insects. Targeted insects included the screwworm fly, several tephritid fruit flies, the gypsy moth, pink bollworm, boll weevil, codling moth and several mosquito species. Many beneficial insects have been, and continue to be, reared for release as biological control measures against pest insects. The most notable of these may be *Trichogramma* species. The glasshouse industry has become heavily dependent on insectary reared insects for pest control. Today, we have an emerging field of insect production as biofactories, both for food and specialty substance sources. These new fields have tremendous potential.

Most agricultural applications of insect rearing require large-scale production. One of the major advancements that has enabled mass production of insects has been the substitution of alternative food sources, and in particular, artificial diets. Artificial diets allow for the mass production of insects in simpler, highly controlled, cost-effective and more convenient ways compared to rearing on native diets such as plants or natural preys. Ideal artificial diets can ultimately serve as suitable substitutes for natural foods and support the production of insects physically and behaviorally, similar to natural diets. Often, the use of artificial diets results in a more uniform and consistent production of high-quality insects for research and field applications. Consequently, the investigation of substitutes for natural food has become a common practice.

The application of new technologies is on the horizon for further refinements in insect production, such as genomics, genetic selection and engineering. Newer multi-omics technologies such as transcriptomics, nutrigenomics and nutrimetabolomics, with increased knowledge of insect microbiome contributions and statistical optimization modeling, have already enabled significant advances in diet formulation. These advances have resulted in a better understanding of the effects of the food stream ingredients on physiological and biochemical functions. Undoubtedly, this will result in improved diet formulations, higher quality control and healthier reared insects with better performance for their targeted applications.

The black soldier fly (*Hermetia illucens* L.) has a global research interest and a growing industrial application since it provides a viable option for countering the environment detriments caused by food waste and a sustainable protein source to feed the growing global population. Hopkins et al. [1] conducted a systematic literature review investigating the impacts of various foodstuffs and substrates for food waste rearing on the protein and amino acid composition of the black soldier fly larvae, finding that plant rearing substrates provide a lower protein content of the total larval mass compared to animal rearing substrates. Pliantiangtam et al. [2] investigated the growth performance, waste reduction efficiency and nutritional composition of the black soldier fly larvae reared on two plant materials (coconut endosperm and soybean curd residue), reporting a similar result of the use of plant rearing sources that yielded lower protein larvae compared with that of animal rearing materials. Furthermore, Lu et al. [3] compared the effects of nine nitrogen sources (i.e., NH_4Cl, $NaNO_3$, urea, uric acid, Gly, L-Glu, L-Glu:L-Asp (1:1, w/w), soybean flour, and fish meal) during food waste larval treatment and characterized the C/N effects on the larval development and bioconversion process. The authors found that organic nitrogen was more suitable than NH_4Cl and $NaNO_3$ as the nitrogen amendment, and an inclusion of small amounts of urea (C/N of 18:1–14:1 and 18:1–16:1) improved the waste reduction performance, and larval protein and lipid bioconversion process, respectively.

The yellow mealworm (*Tenebrio molitor* L.) is another insect species that has been considered as an alternative to fishmeal in animal feed formulations. By utilizing a nutrient self-selection approach, Morales-Ramos et al. [4] demonstrated that the optimum ratio of macro-nutrient intake of this species was 0.06:0.23:0.71 for lipid:protein:carbohydrate. Carbohydrate had positive impacts on food assimilation, food conversion and biomass gain, and several plant materials including cabbage, potato, wheat bran, rice bran (whole and defatted), corn dry distillers' grain, spent brewery dry grain, canola meal and sunflower meal were suitable macro-nutrient components in *T. molitor* diets.

Nikolouli et al. [5] evaluated inactive *Enterobacter* sp. AA26 as a protein source to potentially replace the inactive brewer's yeast, a protein ingredient in the larval diet of the spotted-wing drosophila fly (*Drosophila suzukii* (Matsumura)). This insect species is one of the most damaging insect pests of soft skinned fruits in North America and Europe and is a detrimental invasive pest in South America and Africa. The authors found that *Enterobacter* sp. AA26 provided inadequate nutrition in the larval diet compared with the inactive brewer's yeast. The replacement of *Enterobacter* sp. AA26 with the inactive brewer's yeast resulted in decreases in pupal weight, survival, fecundity and adult recovery.

Improvements in the methodology to store beneficial insects at low temperatures will facilitate biological control programs primarily relying on the mass-release of high-quality bioagents to suppress agricultural pests. Lin et al. [6] characterized impacts of temperatures (4, 7, 10 and 13 °C) and storage durations (10, 15, 20 and 25 days) on the developmental parameters of different pupal age of *Psyttalia incisi* (Silvestri), a dominant parasitoid against *Bactrocera dorsalis* (Hendel) in fruit-producing regions of southern China. The authors reported that the emergence rate of *P. incisi* was significantly affected by storage temperature, storage duration and pupal age interval and their interactions. They further determined the optimum cold storage conditions at a temperature of 13 °C for 10 or 15 d for late-age pupae of this parasitoid. Separately, Lü et al. [7] demonstrated that a temperature of 13 °C was the cold tolerance threshold temperature and the prepupal stage was a critical developmental period for in vitro rearing of another parasitoid *Trichogramma dendrolimi* Matsumura, an important biological control agent of biological control programs.

Insect predators are also important components of biological control programs. Zou et al. [8] investigated the effects of prey species (*Drosophila melanogaster* Meigen or *Bradysia impatiens* (Johannsem)) and prey densities (6–48 preys per predator) on the performance of adult *Coenosia attenuate* Stein, a predator species native to Southern Europe effectively suppressing a wide range of agricultural pests. Their results revealed that *B. impatiens* was the better prey compared with *D. melanogaster*, and the optimal prey density for *C. attenuate* rearing was 12 to 24 adults of *B. impatiens* per predator per day.

Rearing honeybee (*Apis mellifera*) larvae in vitro is an important method for studying bee larvae diseases or the toxicity of pesticides on bees. Kim et al. [9] evaluated the emergence and deformation rates of honeybee (*Apis mellifera ligustica* Spinola) larvae reared in horizontal and vertical positions, finding that a vertical rearing method was the better approach. Compared to the horizontal rearing plates, the vertical rearing plates resulted in a twofold decrease in adult deformation rates and larger adults (11.6 mm vs. 10.8 mm).

Heat-sterilized diets are the key components of high-throughput systems for mass production of insects. Huynh et al. [10] investigated the influence of thermal exposure and lengths of thermal exposure on the quality of a commercialized diet for the western corn rootworm (*Diabrotica virgifera virgifera* LeConte), the most serious pest of maize in the United States and some parts of Europe, to further the goal of developing a diet free of antibiotics and heat-sterilized for this important pest. By using geometric and mathematical approaches, the authors demonstrated non-linear effects of thermal exposure on the performance of diet, whereas no impacts were observed on the exposure intervals evaluated. These findings will guide the continued development of sterilized rootworm diets, facilitating mass production and providing insights into the design of diets for other insects.

This unique topic has been captured in 10 articles that bring together experimental and review papers focusing on different rearing technology approaches to many facets of insect rearing for various purposes. There is every reason to believe that the rapid improvements in insect nutrition and rearing seen over the past decade will be dwarfed by the accomplishments yet to come in the next decade.

Funding: This editorial received no external funding.

Conflicts of Interest: The authors declare no conflict of interest.

References

1. Hopkins, I.; Newman, L.P.; Gill, H.; Danaher, J. The influence of food waste rearing substrates on black soldier fly larvae protein composition: A systematic review. *Insects* **2021**, *12*, 608. [CrossRef] [PubMed]
2. Pliantiangtam, N.; Chundang, P.; Kovitvadhi, A. Growth performance, waste reduction efficiency and nutritional composition of black soldier fly (*Hermetia illucens*) larvae and prepupae reared on coconut endosperm and soybean curd residue with or without supplementation. *Insects* **2021**, *12*, 682. [CrossRef] [PubMed]
3. Lu, Y.; Zhang, S.; Sun, S.; Wu, M.; Bao, Y.; Tong, H.; Ren, M.; Jin, N.; Xu, J.; Zhou, H. Effects of different nitrogen sources and ratios to carbon on larval development and bioconversion efficiency in food waste treatment by black soldier fly larvae (*Hermetia illucens*). *Insects* **2021**, *12*, 507. [CrossRef] [PubMed]
4. Morales-Ramos, J.A.; Rojas, M.G.; Kelstrup, H.C.; Emery, V. Self-selection of agricultural by-products and food ingredients by *Tenebrio molitor* (Coleoptera: Tenebrionidae) and impact on food utilization and nutrient intake. *Insects* **2020**, *11*, 827. [CrossRef] [PubMed]
5. Nikolouli, K.; Sassù, F.; Ntougias, S.; Stauffer, C.; Cáceres, C.; Bourtzis, K. *Enterobacter* sp. AA26 as a protein source in the larval diet of *Drosophila suzukii*. *Insects* **2021**, *12*, 923. [CrossRef]
6. Lin, J.; Yang, D.; Hao, X.; Cai, P.; Guo, Y.; Shi, S.; Liu, C.; Ji, Q. Effect of cold storage on the quality of *Psyttalia incisi* (Hymenoptera: Braconidae), a larval parasitoid of *Bactrocera dorsalis* (Diptera: Tephritidae). *Insects* **2021**, *12*, 558. [CrossRef]
7. Lü, X.; Han, S.-C.; Li, Z.-G.; Li, L.-Y.; Li, J. Gene characterization and enzymatic activities related to trehalose metabolism of in vitro reared *Trichogramma dendrolimi* Matsumura (Hymenoptera: Trichogrammatidae) under sustained cold stress. *Insects* **2020**, *11*, 767. [CrossRef]
8. Zou, D.; Coudron, T.A.; Zhang, L.; Xu, W.; Xu, J.; Wang, M.; Xiao, X.; Wu, H. Effect of prey species and prey densities on the performance of adult *Coenosia attenuata*. *Insects* **2021**, *12*, 669. [CrossRef]
9. Kim, J.; Chon, K.; Kim, B.-S.; Oh, J.-A.; Yoon, C.-Y.; Park, H.-H.; Choi, Y.-S. Horizontal honey-bee larvae rearing plates can increase the deformation rate of newly emerged adult honey bees. *Insects* **2021**, *12*, 603. [CrossRef]
10. Huynh, M.P.; Pereira, A.E.; Geisert, R.W.; Vella, M.G.; Coudron, T.A.; Shelby, K.S.; Hibbard, B.E. Characterization of thermal and time exposure to improve artificial diet for western corn rootworm larvae. *Insects* **2021**, *12*, 783. [CrossRef]

 *insects*

Systematic Review

The Influence of Food Waste Rearing Substrates on Black Soldier Fly Larvae Protein Composition: A Systematic Review

Indee Hopkins , Lisa P. Newman, Harsharn Gill and Jessica Danaher *

School of Science, STEM College, RMIT University, Melbourne 3083, Australia;
s3744095@student.rmit.edu.au (I.H.); lisa.newman@rmit.edu.au (L.P.N.); harsharn.gill@rmit.edu.au (H.G.)
* Correspondence: jessica.danaher@rmit.edu.au; Tel.: +61-3-9925-6117

Simple Summary: The Black Soldier Fly (BSF) is a viable option for countering the environment detriments caused by food waste and can provide a sustainable protein source to feed the growing global population. This systematic literature review investigated the impacts of various foodstuffs and food waste rearing substrates on the protein and amino acid composition of BSF larvae. From the 23 articles included, BSF larvae fed 'Fish waste Sardinella aurita' for two days produced the highest total protein content at 78.8% and rearing substrates 'Fruit and vegetables' reported the lowest protein content at 12.9% of the BSF total mass. However, variation in rearing and analytical methodologies between each study potential undermines the extent to which the rearing substrates may have influenced the overall protein content of BSF larvae.

Citation: Hopkins, I.; Newman, L.P.; Gill, H.; Danaher, J. The Influence of Food Waste Rearing Substrates on Black Soldier Fly Larvae Protein Composition: A Systematic Review. *Insects* **2021**, *12*, 608. https://doi.org/10.3390/insects12070608

Academic Editors: Man P. Huynh, Kent S. Shelby and Thomas A. Coudron

Received: 26 April 2021
Accepted: 29 June 2021
Published: 4 July 2021

Abstract: The Black Soldier Fly (BSF) offers the potential to address two global challenges; the environmental detriments of food waste and the rising demand for protein. Food waste digested by BSF larvae can be converted into biomass, which may then be utilized for the development of value-added products including new food sources for human and animal consumption. A systematic literature search was conducted to identify studies investigating the influence of food waste rearing substrates on BSF larvae protein composition. Of 1712 articles identified, 23 articles were selected for inclusion. Based on the results of this review, BSF larvae reared on 'Fish waste *Sardinella aurita*' for two days reported the highest total protein content at 78.8% and BSF larvae reared on various formulations of 'Fruit and vegetable' reported the lowest protein content at 12.9%. This review is the first to examine the influence of food waste on the protein composition of BSF larvae. Major differences in larval rearing conditions and methods utilized to perform nutritional analyses, potentially influenced the reported protein composition of the BSF larvae. While this review has highlighted the role BSF larvae in food waste management and alternative protein development, their application in nutrition is still in its infancy.

Keywords: alternative protein; amino acid; Black Soldier fly; food waste; insect protein; macronutrients

1. Introduction

With the predicted expansion of the global population expected to reach 8.5 billion by 2030 and 9.7 billion by 2050 [1], we have a vital need to develop and provide a safe and sustainable food system. With an annual increase of 83 million people worldwide, it is estimated that a 70% increase in food production will be required to meet demand, resulting in increased competition for arable land and natural resources such as energy and water [2]. However, our agricultural sectors' ability for growth, particularly in the production of sufficient and quality protein from traditional sources, is constrained by a deficiency of these key resources and presents as a serious challenge. In addition to our current food production system having been deemed as unsustainable from a growth perspective, it is also linked to adverse environmental implications, such as greenhouse gas emissions (GHG) and soil depletion [2].

A simultaneous global issue, despite being paradoxical to our food production and sector growth problem, is that much of the food produced is wasted. Whilst quantification of the extent of the problem is difficult due to lack of consistency in definitions and evaluation methods, it is estimated that as much as one-third of food produced for human consumption is wasted globally each year [3,4]. Food waste negatively impacts the environment in several ways including the net loss of finite resources such as land, water and fuel consumed during food production and distribution. In addition, foodstuffs discarded into landfill are also a contributor to GHG emissions (namely methane), making food waste a growing contributor to climate change [5,6]. It is estimated that food waste is contributing 4.4 gigatons of carbon dioxide (CO_2) emissions into the atmosphere annually [6]. If put into context against national rankings, food waste would be the third-highest contributor of total GHG emissions after that of the United States and China [6]. At present, alternative treatment methods for food waste include incineration, fodder, anaerobic digestion and aerobic composting [7]. However, these methods are either not without their own environmental concerns or unable to be used in isolation to satisfy environmental needs in the long-term.

Another viable alternative to food waste treatment is Conversion of Organic Refuse by Saprophagous (CORS) technologies which use decomposer insects such as the Diptera *Hermetia illucens* L., also known as the Black Soldier Fly (BSF) larvae to manage organic waste. Deteriorated fruits and vegetables, municipal waste, crop waste, and industrial food-processing waste can be quickly and effectively digested by the BSF larvae [8–11]. This process of bioconversion diverts organic food waste from landfill to the production of biomass which can later be utilised for the development of value-added products. The BSF larvae have shown promising results in the production of biodiesel [12], fish feed [13] and have the potential to be used as an alternative source of protein for livestock [14]. The farming of mini livestock (i.e., insects) offers many benefits including high feed conversion ratios, and lower resource inputs in terms of energy, land and water requirements when compared to the farming requirements of traditional livestock (i.e., cattle) [15]. BSF larvae are also safe for human consumption [16], thus offering an opportunity to produce a new sustainable protein source for the growing global population.

Due to the variable nature of the nutritional composition of BSF larvae, the effect of various rearing substrates on BSF development needs to be investigated in further depth. This information can assist in the establishment of efficient growing and production practices in the future. To date, research on the nutritional protein value of BSF larvae fed food waste as the primary rearing substrate has not been synthesised. Thus, the purpose of this systematic literature review is to synthesise and investigate the influence that rearing substrates comprising of foodstuffs or food waste products have on the nutrient composition of BSF larvae.

2. Materials and Methods

2.1. Literature Search Strategy

The current review was performed in adherence to PRISMA-P (Preferred Reporting Items for Systematic review and Meta-Analysis Protocols) guidelines [17]. Scopus, Food Science and Technology Abstracts (FSTA), Web of Science, PubMed, Scifinder and ScienceDirect databases were used to search for articles investigating food waste and the nutritional composition of BSF larvae. The search was limited to articles published between 1 January 2000 and 30 October 2020.

Search terms were pilot tested before commencing the final search to ensure appropriate articles were identified. The final search included keywords searched within three categories (using 'OR') and then combined (using 'AND'). One category searched for articles reporting data on the insect of interest (i.e., Black Soldier Fly, or *Hermetia illucens*). The second category searched for articles reporting on rearing substrates (i.e., food waste, or diets). The third category searched for articles reporting on nutritional outcomes (i.e., protein, amino acids).

2.2. Inclusion and Exclusion Criteria

To be included in the review, articles were required to meet the following criteria: (i) be published as an original research study, with full-text availability in English; (ii) reporting on BSF in the larval or pre-pupae life stage; (iii) including a by-product of food production, foodstuffs or food waste as the rearing substrate for BSF larvae; (iv) reporting an assessment of protein composition of the BSF larvae.

The exclusion criteria included: (i) opinion articles, reviews, narrative reviews and concept papers; (ii) abstract and conference proceedings where a full-text published article could not be obtained; (iii) studies published in a non-English language; (iv) data previously reported elsewhere; (v) studies rearing BSF larvae on alternative waste products not safe for human consumption (i.e., manure); (vi) studies rearing BSF larvae on foodstuffs or food waste products with the inclusion of microbial assistance (i.e., fermentation).

2.3. Study Selection

Articles detected using the search strategy were collated using EndNote. Two reviewers (IH and JD) screened all articles based on titles and abstracts initially, and then by a full-text review. Articles that did not meet the eligibility criteria were excluded. For any articles where it was unclear whether the eligibility criteria were met, full-text articles were obtained, screened and resolved by discussion between three reviewers (IH, LPN and JD) until a consensus was reached.

2.4. Data Extraction and Synthesis

The data extraction method was pilot tested using an article retrieved whilst piloting search terms and refined accordingly. Data from the eligible articles were extracted using a Microsoft Excel spreadsheet, with the following study attributes recorded: year of publication, rearing substrates nutritional and physical characteristics, abiotic factors (including temperature, humidity, and light availability), BSF larvae rearing duration study methodology, statistical analysis, key results and author's conclusions. Data extraction was performed by one reviewer (IH) and verified by secondary reviewers (LPN and JD). In cases of disagreement, a discussion was held until consensus was reached. Extracted data were unable to be combined in a meta-analysis as the studies were not sufficiently homogenous in terms of design and comparator.

2.5. Risk of Bias Assessment

Risk of bias assessment of all included articles was independently undertaken by two reviewers (IH and JD) using the SYRCLE risk of bias tool [18]. The assessment tool considered (i) whether the allocation sequence was adequately generated and applied; (ii) were the groups similar at baseline; (iii) was allocation adequately concealed; (iv) were the animals randomly housed during the experiment; (v) were caregivers and investigators blinded from the knowledge of which intervention each group received; (vi) were larvae selected at random for outcome assessment; (vii) was the outcome assessor blinded; (viii) was incomplete outcome data adequately addressed; (ix) was the study free of selective outcome reporting; (x) was the study apparently free of other issues that could result in high risk of bias. A 'yes' response indicated a low risk of bias, a 'no' response indicated a high risk of bias, and an 'unclear' response indicated that insufficient details had been reported to allow for a valid assessment of the risk of bias [18]. Any discrepancies in the risk of bias assessment were reviewed by a third reviewer (LPN) and resolved by discussion and consensus between the three reviewers.

3. Results

3.1. Search Results

A total of 1712 records were identified through database searches (n = 1051 duplicates), of which 98 were accepted for full-text review after screening by titles and abstracts (Figure 1). Of these, 76 articles were excluded due to the following reasons: BSF larvae was not the primary species investigated (n = 2), rearing substrate did not meet inclusion criteria (n = 17), rearing substrate included added supplementation with microbial assistance or fermentation (n = 2), measurable outcome did not include a relevant nutritional assessment of BSF larvae (n = 33), full-text was not available or was a conference proceeding (n = 10), article was not published in a scientific journal (n = 3) or was not a primary research study (i.e., review article, concept paper or patent application) (n = 9).

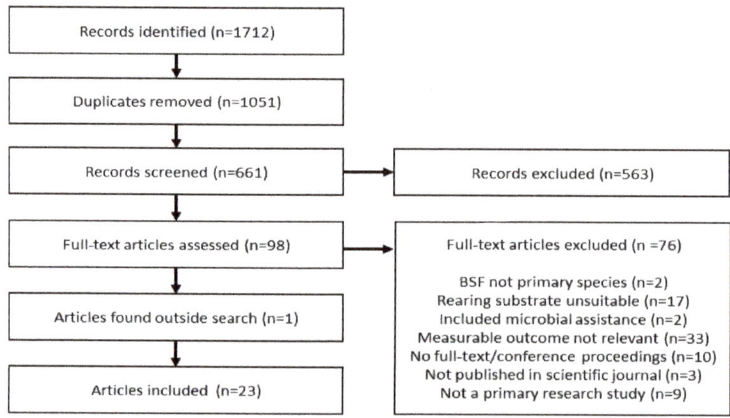

Figure 1. Flow diagram summarising the screening process.

3.2. Rearing Substrates of Black Soldier Fly Larvae

Of the 23 articles included in the current review, 16 articles reported on rearing substrates that contained grain-based ingredients [19–35]. Fifteen articles reported on rearing substrates that contained fruit and vegetable ingredients [19–21,23,25,29–33,35–39]. Six articles reported on rearing substrates that contained animal-based ingredients [24,25,28,39–41], four articles reported on rearing substrates that contained a generic food or kitchen waste description, with no further details regarding included ingredients [24,38,39,41], and one article reported on a rearing substrate that contained seaweed as an ingredient [26].

The rearing duration of the BSF larvae in the experimental trials varied with a reported range between one day [40] and 52 days [37]. Details on larvae feeding frequency were provided by 19 articles, as shown in Table 1 [19–22,24–29,31–34,36–40]. Of these the feeding frequency ranged from a singular feed at the beginning of the experiment [19,35], to a set feeding schedule throughout the experimental trial including daily [26,34,36], weekly [32], and specific days [22,23,25,27,32,33], and ad libitum feeding approaches [21,29,30,37,39,40]. The feed ration provided to BSF larvae was reported by 15 articles [19,20,22–30,33,36,38,39] and varied considerably from 12.2 mg per larvae [23] to 1,530 mg per larvae [24].

Table 1. Rearing substrates (RS) and rearing conditions of the BSF larvae.

Author	Rearing Substrate (RS) (Mixture Ratio)	Rearing Duration (Days)	Feed Ration (mg Per Larvae)	Frequency of Feed	Larval Age at Day 1 (Days)	Temp (°C)	Relative Humidity (%)	Light: Darkness (Hours)
Barbi [19]	RS 1–21: Exotic fruit, pineapple, kiwi, apple, melon (various combinations) RS 22–34: All-year mix, peach, tomato (various combinations) RS 35–49: Legume, corn, pomace, all-year mix (various combinations)	12–47	250–375	-	2–3 instar	27.0 ± 0.5	60.0–70.0	16:8
Barragán-Fonseca [20]	RS 1: Low protein—dried distillers' grains with soluble, cabbage leaves, old bread, cellulose, sunflower oil (unspecified ratio) RS 2: High protein—dried distillers' grains with soluble, cabbage leaves, old bread, cellulose, sunflower oil (unspecified ratio)	21–22	1000	Once at beginning	1	27.0 ± 1.0	70.0 ± 5.0	12:12
Bava [21]	RS 1: Okara RS 2: Maize distillers RS 3: Brewers' grains	16–22	-	Ab libitum	2	25.0 ± 0.5	60.0 ± 0.5	12:12
Cappellozza [36]	RS 1: Fruit and vegetable mix—zucchini, apple, potato, green beans, carrot, pepper, orange, celery, kiwi, plum, eggplant (unspecified ratio)	45	120	Daily after 10 days	Egg	27.0 ± 1.0	50.0 ± 0.5	16:8

Table 1. *Cont.*

Author	Rearing Substrate (RS) (Mixture Ratio)	Rearing Duration (Days)	Feed Ration (mg Per Larvae)	Frequency of Feed	Larval Age at Day 1 (Days)	Temp (°C)	Relative Humidity (%)	Light: Darkness (Hours)
Chia [22]	RS 1: Spent barley RS 2: Spent barley, brewer's yeast RS 3: Spent barley, brewer's yeast, molasses RS 4: Spent malted barley RS 5: Spent malted barley, brewer's yeast RS 6: Spent malted barley, brewer's yeast, molasses RS 7: Spent malted corn RS 8: Spent malted corn, brewer's yeast RS 9: Spent malted corn, brewer's yeast, molasses RS 10: Spent sorghum, barley RS 11: Spent sorghum, barley, brewer's yeast RS 12: Spent sorghum, barley, brewer's yeast, molasses	14–21	≈1000	Every 3 days	Egg	28.0 ± 1.0	70.0 ± 2.0	-
Danieli [23]	RS 1: Control—ground corn, wheat bran, dehydrated alfalfa (5:2:3) RS 2: High non-fibre carbohydrate—ground barley, wheat bran, dehydrated alfalfa (6.8:2:1.2) RS 3: High fibre carbohydrate—ground barley, wheat middlings, dehydrated alfalfa, wheat straw (1.6:5:1:2.4) RS 4: High protein—ground barley, wheat middlings, dehydrated alfalfa (1.5:5.5:3)	21	12.2	Every 2–3 days	6	28.5 ± 0.3	75.6 ± 4.2	-
Ewald [24]	RS 1: Bread RS 2: Fish *O. mykiss*, wheat (5:1) RS 3: Food waste (uncharacterised) RS 4: Fresh mussels *M. edulis* RS 5–9: Bread, fresh mussels *M. edulis* (9:1,8:2,7:3,6:4,5:5)	14	250 230 170 1530 280–300	-	5	28.0	-	-
Gold [25]	RS 1: Mill by-products RS 2: Canteen waste—mix of vegetables with/without dressing, sausage, offal (ratio unspecified) RS 3: Poultry waste RS 4: Vegetable canteen waste—mix of vegetables with/without dressing (ratio unspecified) RS 5: Mixed food waste—(1:1:1 of RS1:RS2:RS3)	9	15–40	Every 3 days	12–14	28.0	70.0	-
Liland [26]	RS 1: Wheat RS 2–10: Wheat, brown algae *A. nodosum* (9:1, 8:2, 7:3, 6:4, 5:5, 4:6, 3:7, 2:8, 1:9) RS 11: Brown algae *A. nodosum*	8	Various as per consumption rates	Daily	8	30.0	65.0	0:24
Liu [27]	RS 1: Brewery by-products (uncharacterised)	15	200	Every 5 days	7	24.5 ± 1.5	40.0 ± 10.0	12:12
Lopes [28]	RS 1: Bread RS 2–4: Bread, fish *O. mykiss* (9.5:0.5,9:1,8.5:1.5)	11–12	250	On days 0,4,7	7	28.0 ± 1.5	45.0 ± 6.3	-
Meneguz [29]	RS 1: Fruit and vegetable mix—celery, oranges, peppers (4.3:2.9:2.8) RS 2: Fruit—apples, oranges, apple leftovers, strawberries, mandarins, pears, kiwis, bananas, lemons (4.8:1.5:1.4:0.7:0.5:0.4:0.3:0.2:0.2) RS 3: Winery by-products—grape seeds, pulp, skins, stems, leaves (ratio unspecified) RS 4: Brewery by-products—barley brewers' grains	23–28	1000	Ad libitum	6	27.0 ± 5.0	70.0 ± 5.0	24:0
Oonincx [30]	RS 1: High protein high fat—spent grains, beer yeast, cookie remains (6:2:2) RS 2: High protein low fat—beer yeast, potato stem peelings, beet molasses (5:3:2) RS 3: Low protein high fat—cookie remains, bread (5:5) RS 4: Low protein low fat—potato steam peelings, beet molasses, bread (3:2:5)	21–37	40	Ad libitum	<1	28.0	70.0	12:12
Salomone [31]	RS 1: Food waste—vegetable, meat/fish, bread/pasta/rice, other (6.5:0.5:2.5:0.5)	12	-	-	-	30–35	>65.0	-
Shumo [32]	RS 1: Kitchen waste—potato peelings, carrot, rice, bread debris (ratio unspecified) RS 2: Brewery by-product spent grain	21–28	-	Weekly	Neonate	28.0 ± 2.0	65.0 ± 5.0	-
Spranghers [33]	RS 1: Restaurant waste—potato, rice, pasta, vegetables (ratio unspecified)	15–19	600	Every 3 days	6–8	27.0 ± 1.0	65.0 ± 5.0	-
Tinder [34]	RS 1: Sorghum RS 2–4: Sorghum, cowpea (7.5:2.5, 5:5, 2.5:7.5) RS 5: Cowpea	28–38	33.3	Daily	4	28.0 ± 2.0	70.0	14:10
Tschirner [35]	RS 1: Carbohydrate—wheat middlings RS 2: Protein—dried distillers' grains with soluble RS 3: Fibre—sugar beet	15	-	Once at beginning	8	28.0	73.8–75.7	-
Cappellozza [36]	RS 1: Fruit and vegetable mix—zucchini, apple, potato, green beans, carrot, pepper, orange, celery, kiwi, plum, eggplant (unspecified ratio)	45	120	Daily after 10 days	Egg	27.0 ± 1.0	50.0 ± 0.5	16:8
Jucker [37]	RS 1: Fruit—apple, pear, orange (3.3:3.3:3.3) RS 2: Vegetable—lettuce, string green beans, cabbage (3.3:3.3:3.3) RS 3: Fruit and vegetable mix—(1:1 of RS1:RS2)	37–52	-	Ad libitum	9	25.0 ± 0.5	60.0 ± 0.5	12:12
Lalander [38]	RS 1: Food waste (uncharacterised) RS 2: Fruit and vegetable mix—lettuce, apple, potato (5:3:2)	19–47	40	Every 2–3 days	10	28.0	-	-
Nguyen [39]	RS 1: Kitchen waste (animal and plant matter) RS 2: Fruit and vegetable (uncharacterised) RS 3: Fish (uncharacterised)	19–40	40	Ad libitum	4	28.0	60.0 ± 10.0	-
Barroso [40]	RS 1: Fish *S. aurita*	1–12	-	Ab libitum	3	26.0 ± 1.0	65.0 ± 5.0	-
Surendra [41]	RS 1: Food waste (uncharacterised)	-	-	-	-	-	-	-

≈, approximately; g, grams; mg, milligrams; C, Celsius. Nguyen et al. [42] article data used to support Nguyen et al. [39] reporting of rearing conditions. Second or third stage instar implies ≈ 15 days of age [43]. Neonate implies <5 days of age [44].

3.3. Rearing Abiotic Conditions of Black Soldier Fly Larvae

Abiotic conditions of the 23 articles included in the current review are shown in Table 1. Of these, ten articles included details regarding duration of light and dark exposure hours of BSF larvae [19–21,25–27,29,30,34,36,37]. Five articles reporting a duration of 12 h of light and 12 h of darkness [20,21,27,30,37], two articles reporting 16 h of light and eight hours of darkness [19,36], one article reporting 24 h of light [29], one article reporting 24 h darkness [25] and one article reporting 14 h of light and ten hours of darkness [34]. Twenty articles included in this review provided data on the relative humidity of the BSF larvae rearing environment [19–40], with ranges varying from 40.0% relative humidity [27] to 75.6% relative humidity [23].

Twenty-two articles included information regarding the temperature of the rearing environment [19–40], with a range of 24.5 °C [27] to 32.5 °C [31]. The age of larvae at the beginning of the experimental trials was reported in 21 articles [19–29,31–40], with BSF larvae ages ranging from eggs which were directly inoculated onto the rearing substrate [22,36] to BSF larvae aged 14 days before being introduced to the substrate [25]. The rearing duration of BSF larvae was reported in 20 articles [19–21,23–30,32–40], with rearing duration ranging from eight days [26,29] to 52 days [37].

3.4. Macronutrient Composition of Rearing Substrates

Twenty of the 23 articles included in this review provided details on the total protein composition of the rearing substrate provided to the BSF larvae (Table 2) [19–41]. Eleven articles reported details of the rearing substrate total lipid content [19,24–30,33,34,37,39]. Thirteen articles additionally reported details of the rearing substrate total carbohydrate content [19–30,33,34,37,39].

Table 2. Macronutrient Composition of Rearing Substrates (RS).

Author	Rearing Substrate (RS) (Mixture Ratio)	Protein (%) DM	Protein (%) FW	Lipid (%) DM	Lipid (%) FW	Carbohydrate (%) DM	Carbohydrate (%) FW
Barbi [19]	RS 1: Exotic fruit	0.7	-	0.1	-	15.4	-
	RS 2: Pineapple	0.5	-	0.2	-	11.4	-
	RS 3: Kiwi	0.9	-	0.1	-	7.4	-
	RS 4: Apple	0.4	-	0.0	-	6.9	-
	RS 5: Melon	0.5	-	0.1	-	0.0	-
	RS 6: Tomato	3.0	-	0.4	-	0.0	-
	RS 7: Peach	0.9	-	0.8	-	4.3	-
	RS 8: Pomace	2.0	-	3.7	-	8.2	-
	RS 9: Legume	17.0	-	0.7	-	6.3	-
	RS 10: Corn	11.5	-	1.4	-	0.3	-
Barragán–Fonseca [20]	RS 1: Low protein—dried distillers' grains with soluble, cabbage leaves, old bread, cellulose and sunflower oil (ratio unspecified)	10.0	-	-	-	30.0	-
	RS 2: High protein—dried distillers' grains with soluble, cabbage leaves, old bread, cellulose and sunflower oil (ratio unspecified)	17.0	-	-	-	30.0	-
Bava ~ [21]	RS 1: Maize distillers	39.2	-	-	-	7.5	-
	RS 2: Okara	29.5	-	-	-	17.3	-
	RS 3: Brewer's grains	15.8	-	-	-	11.2	-
Chia [22]	RS 1: Spent barley	30.3	-	6.4	-	-	-
	RS 2: Spent barley, brewer's yeast (3.6:6.4)	32.0	-	5.4	-	-	-
	RS 3: Spent barley, brewer's yeast, molasses (3.6:3.2:3:2)	22.1	-	4.0	-	-	-
	RS 4: Spent malted barley	28.9	-	6.8	-	-	-
	RS 5: Spent malted barley, brewer's yeast (3.6:6.4)	30.2	-	7.0	-	-	-
	RS 6: Spent malted barley, brewer's yeast, molasses (3.6:3.2:3:2)	22.3	-	3.2	-	-	-
	RS 7: Spent malted corn	27.4	-	6.5	-	-	-
	RS 8: Spent malted corn, brewer's yeast (3.6:6.4)	27.7	-	6.0	-	-	-
	RS 9: Spent malted corn, brewer's yeast, molasses (3.6:3.2:3:2)	19.1	-	3.4	-	-	-
	RS 10: Spent sorghum, barley	29.4	-	11.8	-	-	-
	RS 11: Spent sorghum, barley, brewer's yeast (3.6:6.4)	31.4	-	9.5	-	-	-
	RS 12: Spent sorghum, barley, brewer's yeast, molasses (3.6:3.2:3:2)	21.7	-	5.2	-	-	-
Danieli ^ [23]	RS 1: Control—ground corn, wheat bran, dehydrated alfalfa (5:2:3)	10.6	-	-	-	59.7	-
	RS 2: High non-fibre carbohydrate—ground barley, wheat bran, dehydrated alfalfa (6.8:2:1.2)	11.1	-	-	-	68.9	-
	RS 3: High fibre carbohydrate—ground barley, wheat middlings, dehydrated alfalfa, wheat straw (1.6:5:1:2.4)	11.2	-	-	-	51.1	-
	RS 4: High protein—ground barley, wheat middlings, dehydrated alfalfa (1.5:5.5:3)	13.8	-	-	-	56.1	-

Table 2. *Cont.*

| Author | Rearing Substrate (RS) (Mixture Ratio) | Rearing Substrate (RS) | | | | | |
| | | Protein (%) | | Lipid (%) | | Carbohydrate (%) | |
		DM	FW	DM	FW	DM	FW
Ewald [24]	RS 1: Bread	13.5	-	5.3	-	78.6	-
	RS 2: Fish *O. mykiss*, wheat (5:1)	41.8	-	22.5	-	27.8	-
	RS 3: Food waste (uncharacterised)	20.5	-	20.7	-	48.4	-
	RS 4: Fresh mussels *M. edulis*	17.4	-	1.4	-	7.2	-
	RS 5: Bread, fresh mussels *M. edulis* (9:1)	14.5	-	3.0	-	56.9	-
	RS 6: Bread, fresh mussels *M. edulis* (8:2)	15.3	-	2.3	-	42.9	-
	RS 7: Bread, fresh mussels *M. edulis* (7:3)	15.8	-	1.9	-	33.7	-
	RS 8: Bread, fresh mussels *M. edulis* (6:4)	16.2	-	1.6	-	27.1	-
	RS 9: Bread, fresh mussels *M. edulis* (5:5)	16.4	-	1.3	-	22.2	-
Gold [25]	RS 1: Milled by-products	14.5	-	3.0	-	23.3	-
	RS 2: Canteen waste—mix of vegetables with/without dressing, sausage, offal (ratio unspecified)	32.2	-	34.9	-	7.5	-
	RS 3: Poultry waste	37.3	-	42.9	-	0.3	-
	RS 4: Vegetable canteen waste—mix of vegetables with/without dressing (ratio unspecified)	12.1	-	28.9	-	15.5	-
	RS 5: Mixed food waste—(1:1:1 of RS1:RS2:RS3)	19.6	-	22.3	-	15.4	-
Liland [26]	RS 1: Wheat	10.8	-	4.8	-	-	-
	RS 2: Wheat, brown algae *A. nodosum* (9:1)	9.8	-	4.8	-	-	-
	RS 3: Wheat, brown algae *A. nodosum* (8:2)	9.6	-	4.2	-	-	-
	RS 4: Wheat, brown algae *A. nodosum* (7:3)	8.6	-	3.3	-	-	-
	RS 5: Wheat, brown algae *A. nodosum* (6:4)	8.5	-	3.3	-	-	-
	RS 6: Wheat, brown algae *A. nodosum* (5:5)	7.4	-	3.3	-	-	-
	RS 7: Wheat, brown algae *A. nodosum* (4:6)	6.7	-	4.0	-	-	-
	RS 8: Wheat, brown algae *A. nodosum* (3:7)	6.5	-	2.8	-	-	-
	RS 9: Wheat, brown algae *A. nodosum* (2:8)	5.1	-	2.1	-	-	-
	RS 10: Wheat, brown algae *A. nodosum* (1:9)	5.3	-	2.4	-	-	-
	RS 11: Brown algae *A. nodosum*	4.5	-	2.0	-	-	-
Liu* [27]	RS 1: Brewery by-product (uncharacterised)	22.6	-	5.8	-	8.9	-
Lopes [28]	RS 1: Bread waste	8.2	-	0.0	-	46.1	-
	RS 2: Fish waste *O. mykiss*	60.3	-	32.5	-	0.0	-
Meneguz ^ [29]	RS 1: Fruit and vegetable mix—celery, oranges, peppers (4.3:2.9:2.8)	12.0	-	2.1	-	58.5	-
	RS 2: Fruit-apples, oranges, apple leftovers, strawberries, mandarins, pears, kiwis, bananas, lemons (4.8:1.5:1.4:0.7:0.5:0.4:0.3:0.2:0.2)	4.6	-	1.0	-	75.7	-
	RS 3: Winery by-products—grape seeds, pulp, skins, stems, leaves (ratio unspecified)	11.7	-	7.4	-	13.4	-
	RS 4: Brewery by-products—barley brewers' grains	20.1	-	8.2	-	22.6	-
Oonincx [30]	RS 1: High protein high fat—spent grains, beer yeast, cookie remain (6:2:2)	21.9	-	8.9	-	-	-
	RS 2: High protein low fat—beer yeast, potato steam peelings, beet molasses (5:3:2)	22.9	-	0.4	-	-	-
	RS 3: Low protein high fat—cookie remains, bread (5:5)	12.9	-	14.0	-	-	-
	RS 4: Low protein low fat—potato steam peelings, beet molasses, bread (3:2:5)	14.4	-	1.5	-	-	-
Salomone [31]	RS 1: Food waste—vegetable, meat/fish, bread/pasta/rice, other (6.5:0.5:2.5:0.5)	-	-	-	-	-	-
Shumo [32]	RS 1: Kitchen waste—potato peelings, carrot, rice, bread debris (ratio unspecified)	20.0	-	-	-	-	-
	RS 2: Brewery by-product spent grain	12.2	-	-	-	-	-
Spranghers ^ [33]	RS 1: Restaurant waste—potato, rice, pasta, vegetables (ratio unspecified)	15.7	-	-	-	61.8	-
Tinder *~ [34]	RS 1: Sorghum	3.5	-	1.0	-	23.7	-
	RS 2: Cowpea	7.7	-	0.5	-	20.8	-
Tschirner [35]	RS 1: Carbohydrate—wheat middlings	22.0	-	-	-	-	-
	RS 2: Protein—dried distillers' grains with soluble	31.2	-	-	-	-	-
	RS 3: Fibre—sugar beet	8.5	-	-	-	-	-
Cappellozza [36]	RS 1: Fruit and vegetable mix—zucchini, apple, potato, green beans, carrot, pepper, orange, celery, kiwi, plum, eggplant (unspecified ratio)	-	-	-	-	-	-
Jucker *~ [37]	RS 1: Fruit—apple, pear, orange (3.3:3.3:3.3)	-	0.4	-	0.1	-	8.9
	RS 2: Vegetable—lettuce, string green beans, cabbage (3.3:3.3:3.3)	-	2.0	-	0.2	-	2.4
	RS 3: Fruit and vegetable mix—(1:1 of RS1:RS2)	-	1.2	-	0.2	-	5.6
Lalander [38]	RS 1: Food waste (uncharacterised)	22.2	-	-	-	55.0 ~	-
	RS 2: Fruit and vegetable mix—lettuce, apple, potato (5:3:2)	13.2	-	-	-	72.6 ~	-
Nguyen * [39]	RS 1: Kitchen waste (animal and plant matter)	20.4	-	19.6	-	56.8	-
	RS 2: Fruit and vegetable (uncharacterised)	20.1	-	1.6	-	69.0	-
	RS 3: Fish (uncharacterised)	50.0	-	36.2	-	0.6	-
Barroso [40]	RS 1: Fish waste *S. aurita*	72.7	-	-	-	-	-
Surendra [41]	RS 1: Food waste (uncharacterised)	-	-	-	-	-	-

Results presented as a percentage. * Indicative of original article presenting data as g/100 g of rearing substrate. ^ indicative of original article presenting data as g/kg of rearing substrate. ~ indicative of original article nutrient data acquired from database or literature. Dashes used to indicate where data is unreported in the original article. Chia et al. [45] article data used to support Chia et al. [22] reporting of rearing substrate composition. DM dry matter, FW fresh weight.

There were substantial differences in the macronutrient composition of the rearing substrate provided, with an average total composition of 17.5% total protein, 7.2% total lipid and 27.3% total carbohydrate, when reported on a dry matter basis. Rearing substrate 'Apple' showed the lowest total protein content of 0.4% of dry matter [19]. The highest total protein content of a rearing substrate was reported for 'Fish waste *S. aurita*', with 72.7% protein dry matter [40]. Rearing substrates 'Apple' [19] and 'Bread waste' [28] both showed the lowest total lipid content of 0.0% of dry matter. The highest total lipid content in a substrate was reported for 'Poultry waste', with 42.9% lipid dry matter [25]. Rearing substrates 'Melon', 'Tomato' [19] and 'Fish waste *O. mykiss*' [28] all showed the lowest total carbohydrate content of 0.0% of dry matter. The highest total carbohydrate content of a rearing substrate was reported for 'Bread', with 78.6% carbohydrate dry matter [24].

Only one article reported rearing substrate macronutrient composition on a fresh weight basis, finding the highest protein and lipid content in substrate 'Vegetable—lettuce, string green beans, cabbage (ratio of 3.3:3.3:3.3)' (2.0% and 0.2% of fresh weight, respectively) and the highest substrate carbohydrate content in substrate 'Fruit—apple, pear, orange (ratio of 3.3:3.3:3.3)' (8.9% of fresh weight) [37].

3.5. Amino Acid Composition of Rearing Substrates

Two articles reported the amino acid content of the rearing substrate provided to the BSF larvae (Supplementary Table S1) [26,33]. Results collected indicated that glutamate was the most abundant amino acid found, accounting for 25.9% of total protein dry matter in 'Brown algae *A. nodosum*' [26]. No other substrate composition included this amino acid in isolation.

Cysteine and tryptophan were the least abundant amino acids reported in the rearing substrate with 'Restaurant waste—potato, rice, pasta, vegetable (ratio unspecified)' showing the presence of both amino acids at 0.2% of total protein dry matter [33].

3.6. Macronutrient Composition of Black Soldier Fly Larvae Reared on Food Waste

As per the inclusion criteria, all articles included in this review provided details on the protein composition of BSF larvae reared on food waste substrates. Of these, 22 articles reported BSF larvae protein composition on a dry matter basis, with one article reporting on a fresh weight basis (Table 3).

Table 3. Macronutrient Composition of Black Solider Fly Larvae Reared on Food Waste.

Author	Rearing Substrate (RS) (Mixture Ratio)	Black Soldier Fly Larvae Protein (%)		Lipid (%)		Carbohydrate (%)	
		DM	FW	DM	FW	DM	FW
	RS 1: Exotic fruit, melon (5:5)	14.0	-	6.5	-	-	-
	RS 2: Exotic fruit, pineapple, kiwi, apple, melon (1:1:6:1:1)	13.5	-	7.7	-	-	-
	RS 3: Pineapple	13.4	-	5.0	-	-	-
	RS 4: Melon	15.1	-	5.1	-	-	-
	RS 5: Apple	13.9	-	7.0	-	-	-
	RS 6: Exotic fruit	13.8	-	9.6	-	-	-
	RS 7: Exotic fruit, pineapple, kiwi, apple, melon (2:2:2:2:2)	12.9	-	8.8	-	-	-
	RS 8: Exotic fruit, kiwi (5:5)	12.9	-	8.4	-	-	-
	RS 9: Pineapple, melon (5:5)	13.9	-	6.0	-	-	-
	RS 10: Kiwi, melon (5:5)	13.5	-	6.0	-	-	-
	RS 11: Pineapple, apple (5:5)	13.9	-	7.7	-	-	-
	RS 12: Exotic fruit, pineapple, kiwi, apple, melon (1:1:1:6:1)	14.2	-	7.3	-	-	-
	RS 13: Apple, melon (5:5)	14.3	-	7.5	-	-	-
	RS 14: Exotic fruit, pineapple, kiwi, apple, melon (1:1:1:1:6)	13.9	-	5.8	-	-	-
	RS 15: Pineapple, kiwi (5:5)	14.5	-	7.6	-	-	-
	RS 16: Exotic fruit, pineapple (5:5)	14.1	-	8.8	-	-	-
	RS 17: Kiwi	13.1	-	7.9	-	-	-
	RS 18: Exotic fruit, apple (5:5)	13.9	-	9.5	-	-	-
	RS 19: Exotic fruit, pineapple, kiwi, apple, melon (6:1:1:1:1)	13.6	-	8.5	-	-	-
Barbi [19]	RS 20: Exotic fruit, pineapple, kiwi, apple, melon (1:6:1:1:1)	14.2	-	7.5	-	-	-
	RS 21: Kiwi, apple (5:5)	13.1	-	7.0	-	-	-
	RS 22: Peach, tomato (6.7:3.3)	15.7	-	13.7	-	-	-
	RS 23: Peach	15.5	-	13.6	-	-	-
	RS 24: All-year mix, peach, tomato (6.7:1.6:1.7)	15.0	-	12.1	-	-	-
	RS 25: All-year mix, peach, tomato (3.4:3.3:3.3)	15.3	-	12.1	-	-	-
	RS 26: All-year mix, tomato (3.3:6.7)	15.6	-	10.6	-	-	-
	RS 27: All-year mix, peach (6.7:3.3)	14.6	-	10.7	-	-	-
	RS 28: All-year mix, peach, tomato (1.7:6.7:1.6)	16.4	-	14.2	-	-	-
	RS 29: All-year mix, tomato (6.7:3.3)	15.3	-	7.5	-	-	-

Table 3. *Cont.*

| Author | Rearing Substrate (RS) (Mixture Ratio) | Black Soldier Fly Larvae | | | | | |
| | | Protein (%) | | Lipid (%) | | Carbohydrate (%) | |
		DM	FW	DM	FW	DM	FW
	RS 30: All-year mix	14.8	-	9.0	-	-	-
	RS 31: Peach, tomato (3.3:6.7)	14.8	-	12.2	-	-	-
	RS 32: All-year mix, peach, tomato (1.6:1.7:6.7)	15.9	-	12.6	-	-	-
	RS 33: Tomato	15.7	-	11.4	-	-	-
	RS 34: All-year mix, peach (3.3:6.7)	16.0	-	12.5	-	-	-
	RS 35: Legume, corn, pomace, all-year mix (1.25:6.25:1.25:1.25)	16.9	-	12.0	-	-	-
	RS 36: Corn, all-year mix (5:5)	16.9	-	11.6	-	-	-
	RS 37: Corn	18.4	-	10.2	-	-	-
	RS 38: Legume, corn, pomace, all-year mix (1.25:1.25:6.25:1.25)	16.5	-	8.7	-	-	-
	RS 39: Legume, corn, pomace, all-year mix (6.25:1.25:1.25:1.25)	16.2	-	11.0	-	-	-
	RS 40: Legume	17.3	-	7.0	-	-	-
	RS 41: Legume, pomace (5:5)	16.5	-	6.2	-	-	-
	RS 42: All-year mix	15.3	-	8.9	-	-	-
	RS 43: Legume, corn (5:5)	17.6	-	7.4	-	-	-
	RS 44: Pomace, all-year mix (5:5)	14.8	-	4.9	-	-	-
	RS 45: Legume, corn, pomace, all-year mix (1.25:1.25:1.25:6.25)	17.8	-	7.6	-	-	-
	RS 46: Pomace	14.8	-	5.6	-	-	-
	RS 47: Corn, pomace (5:5)	17.4	-	7.8	-	-	-
	RS 48: Legume, all-year mix (5:5)	16.9	-	7.7	-	-	-
	RS 49: Legume, corn, pomace, all-year mix (2.5:2.5:2.5:2.5)	17.7	-	8.7	-	-	-
Barragán-Fonseca [20]	RS 1: Low protein—dried distillers' grains with soluble, cabbage leaves, old bread, cellulose, sunflower oil (ratio unspecified)	47.0	-	20.0	-	-	-
	RS 2: High protein—dried distillers' grains with soluble, cabbage leaves, old bread, cellulose, sunflower oil (ratio unspecified)	46.0	-	32.0	-	-	-
Bava [21]	RS 1: Okara	51.2	-	-	-	-	-
	RS 2: Maize distillers	53.4	-	-	-	-	-
	RS 3: Brewers' grains (uncharacterised)	54.1	-	-	-	-	-
Chia [22]	RS 1: Spent barley	37.4	-	33.2	-	-	-
	RS 2: Spent barley, brewer's yeast	41.9	-	22.5	-	-	-
	RS 3: Spent barley, brewer's yeast, molasses	31.7	-	49.0	-	-	-
	RS 4: Spent malted barley	39.9	-	21.1	-	-	-
	RS 5: Spent malted barley, brewer's yeast	41.3	-	17.1	-	-	-
	RS 6: Spent malted barley, brewer's yeast, molasses	29.9	-	39.3	-	-	-
	RS 7: Spent malted corn	40.6	-	25.5	-	-	-
	RS 8: Spent malted corn, brewer's yeast	39.8	-	21.1	-	-	-
	RS 9: Spent malted corn, brewer's yeast, molasses	31.8	-	42.3	-	-	-
	RS 10: Spent sorghum, barley	40.3	-	29.7	-	-	-
	RS 11: Spent sorghum, barley, brewer's yeast	45.7	-	9.5	-	-	-
	RS 12: Spent sorghum, barley, brewer's yeast, molasses	44.6	-	11.4	-	-	-
Danieli [23]	RS 1: Control—ground corn, wheat bran, dehydrated alfalfa (5:2:3)	34.0	-	-	-	-	-
	RS 2: High non-fibre carbohydrate—ground barley, wheat bran, dehydrated alfalfa (6.8:2:1.2)	22.2	-	-	-	-	-
	RS 3: High fibre carbohydrate—ground barley, wheat middlings, dehydrated alfalfa, wheat straw (1.6:5:1:2.4)	34.7	-	-	-	-	-
	RS 4: High protein—ground barley, wheat middlings, dehydrated alfalfa (1.5:5.5:3)	34.2	-	-	-	-	-
Ewald [24]	RS 1: Bread	39.2	-	57.8	-	-	-
	RS 2: Fish *O. mykiss*, wheat (5:1)	52.6	-	46.7	-	-	-
	RS 3: Food waste (uncharacterised)	36.6	-	40.7	-	-	-
	RS 4: Fresh mussels *M. edulis*	44.6	-	33.1	-	-	-
	RS 5: Bread, fresh mussels *M. edulis* (9:1)	32.8	-	20.4	-	-	-
	RS 6: Bread, fresh mussels *M. edulis* (8:2)	34.2	-	19.6	-	-	-
	RS 7: Bread, fresh mussels *M. edulis* (7:3)	33.8	-	17.9	-	-	-
	RS 8: Bread, fresh mussels *M. edulis* (6:4)	36.1	-	17.9	-	-	-
	RS 9: Bread, fresh mussels *M. edulis* (5:5)	37.9	-	16.1	-	-	-
Gold [25]	RS 1: Mill by-products (uncharacterised)	42.1	-	-	-	-	-
	RS 2: Canteen waste—mix vegetables with/without dressing, sausage, offal (ratio unspecified)	36.1	-	-	-	-	-
	RS 3: Poultry waste	31.5	-	-	-	-	-
	RS 4: Vegetable canteen waste—mix vegetables with/without dressing	24.5	-	-	-	-	-
	RS 5: Mixed food waste—(1:1:1 of RS1:RS2:RS3)	28.6	-	-	-	-	-
Liland [26]	RS 1: Wheat	40.0	-	33.8	-	-	-
	RS 2: Wheat, brown algae *A. nodosum* (9:1)	37.9	-	-	-	-	-
	RS 3: Wheat, brown algae *A. nodosum* (8:2)	35.9	-	-	-	-	-
	RS 4: Wheat, brown algae *A. nodosum* (7:3)	35.3	-	-	-	-	-
	RS 5: Wheat, brown algae *A. nodosum* (6:4)	33.5	-	-	-	-	-
	RS 6: Wheat, brown algae *A. nodosum* (5:5)	33.7	-	22.2	-	-	-
	RS 7: Wheat, brown algae *A. nodosum* (4:6)	37.4	-	-	-	-	-
	RS 8: Wheat, brown algae *A. nodosum* (3:7)	42.3	-	-	-	-	-
	RS 9: Wheat, brown algae *A. nodosum* (2:8)	41.0	-	-	-	-	-
	RS 10: Wheat, brown algae *A. nodosum* (1:9)	39.3	-	-	-	-	-
	RS 11: Brown algae *A. nodosum*	41.3	-	8.1	-	-	-
Liu * [27]	RS 1: Brewery by-product (uncharacterised)	49.9	-	33.7	-	-	-
Lopes [28]	RS 1: Bread	28.0	-	-	-	-	-
	RS 2: Bread, fish *O. mykiss*, (9.5:0.5)	39.1	-	-	-	-	-
	RS 3: Bread, fish *O. mykiss*, (9:1)	42.7	-	-	-	-	-
	RS 4: Bread, fish *O. mykiss*, (8.5:1.5)	44.8	-	-	-	-	-

Table 3. *Cont.*

Author	Rearing Substrate (RS) (Mixture Ratio)	Black Soldier Fly Larvae					
		Protein (%)		Lipid (%)		Carbohydrate (%)	
		DM	FW	DM	FW	DM	FW
Meneguz [29]	RS 1: Fruit and vegetable mix—celery, oranges, peppers (4.3:2.9:2.8)	41.9	-	25.3	-	-	-
	RS 2: Fruit—apples, oranges, apple leftovers, strawberries, mandarins, pears, kiwis, bananas, lemons (4.8:1.5:1.4:0.7:0.5:0.4:0.3:0.2:0.2)	30.8	-	39.8	-	-	-
	RS 3: Winery by-products—grape seeds, pulp, skins, stems, leaves (unspecified ratio)	34.4	-	28.7	-	-	-
	RS 4: Brewery by-products—barley brewers' grains	53.0	-	28.3	-	-	-
Oonincx [30]	RS 1: High protein high fat—spent grains, beer yeast, cookie remain (6:2:2)	46.3	-	24.1	-	-	-
	RS 2: High protein low fat—beer yeast, potato steam peelings, beet molasses (5:3:2)	43.5	-	24.9	-	-	-
	RS 3: Low protein high fat—cookie remains, bread (5:5)	38.8	-	27.4	-	-	-
	RS 4: Low protein low fat—potato steam peelings, beet molasses, bread (3:2:5)	38.3	-	32.9	-	-	-
Salomone [31]	RS 1: Food waste—vegetable, meat/fish, bread/pasta/rice, other (6.5:0.5:2.5:0.5)	48.0	-	35.0	-	-	-
Shumo [32]	RS 1: Kitchen waste—potato peelings, carrot, rice, bread debris (ratio unspecified)	33.0	-	-	-	-	-
	RS 2: Brewery by-product spent grain	41.3	-	-	-	-	-
Spranghers [33]	RS 1: Restaurant waste—potato, rice, pasta, vegetables (ratio unspecified)	43.1	-	-	-	-	-
Tinder [34]	RS 1: Sorghum	44.1	-	-	-	-	-
	RS 2: Sorghum, cowpea (7.5:2.5)	44.9	-	-	-	-	-
	RS 3: Sorghum, cowpea (5:5)	45.4	-	-	-	-	-
	RS 4: Sorghum, cowpea (2.5:7.5)	46.1	-	-	-	-	-
	RS 5: Cowpea	47.3	-	-	-	-	-
Tschirner [35]	RS 1: Carbohydrate—wheat middlings	37.2	-	-	-	-	-
	RS 2: Protein—dried distillers' grains with soluble	44.6	-	-	-	-	-
	RS 3: Fibre—sugar beet	52.3	-	-	-	-	-
Cappellozza [36]	RS 1: Fruit and vegetable mix—zucchini, apple, potato, green beans, carrot, pepper, orange, celery, kiwi, plum, eggplant (unspecified ratio)	39.4	-	35.6	-	-	-
Jucker * [37]	RS 1: Fruit—apple, pear, orange (3.3:3.3:3.3)	-	11.7	-	21.0 $\approx$	-	-
	RS 2: Vegetable—lettuce, string green beans, cabbage (3.3:3.3:3.3)	-	13.2	-	2.0 $\approx$	-	-
	RS 3: Fruit and vegetable mix—(1:1 of RS1:RS2)	-	17.6	-	12.0 $\approx$	-	-
Lalander [38]	RS 1: Food waste (uncharacterised)	39.2	-	-	-	-	-
	RS 2: Fruit and vegetable mix-lettuce, apple, potato (5:3:2)	41.3	-	-	-	-	-
Nguyen * [39]	RS 1: Kitchen waste (animal and plant matter)	21.2	-	-	-	-	-
	RS 2: Fruit and vegetable (uncharacterised)	12.9	-	2.2	-	8.4	-
	RS 3: Fish (uncharacterised)	19.4	-	11.6	-	12.7	-
Barroso [40]	RS 1: Fish waste *S. aurita*-reared 1 day	77.4	-	-	-	-	-
	RS 2: Fish waste *S. aurita*-reared 2 days	78.8	-	-	-	-	-
	RS 3: Fish waste *S. aurita*-reared 4 days	75.4	-	-	-	-	-
	RS 4: Fish waste *S. aurita*-reared 6 days	50.6	-	-	-	-	-
	RS 5: Fish waste *S. aurita*-reared 8 days	71.3	-	-	-	-	-
	RS 6: Fish waste *S. aurita*-reared 10 days	61.5	-	-	-	-	-
	RS 7: Fish waste *S. aurita*-reared 12 days	61.8	-	-	-	-	-
Surendra [41]	RS 1: Food waste (uncharacterised)	43.7	-	31.8	-	12.3	-

Results presented as a percentage of total BSF larvae biomass unless indicated. Liland et al. [26] presented as total sum of amino acids. $\approx$ approximate figure; * indicative of original article presenting data as g/100g BSF larvae biomass, ^ indicative of original article presenting data as g/kg BSF larvae biomass. Dashes used to indicate where data is unreported in the original article, DM dry matter, FW fresh weight.

There was an average of 31.2% total protein when reported on a dry matter basis. Rearing substrate 'Exotic fruit, pineapple, kiwi, apple, melon (ratio of 2:2:2:2:2)' for 26.7 days, 'Exotic fruit, kiwi (ratio of 5:5)' for 25.3 days [19] and 'Fruit and vegetable (uncharacterised)' for an unspecified number of days [39], equally resulted in the lowest total protein content of 12.9% of BSF larvae dry matter. The highest BSF larvae total protein content was reported for larvae reared for two days on 'Fish waste *S. aurita*' with 78.8% total protein BSF larvae dry matter [40]. This was followed by 'Fish waste *S. aurita*' reared for one and four days and resulting in 77.4% and 75.4% total protein of BSF larvae dry matter, respectively [40].

Twelve articles reported additional details of the BSF larvae total lipid content on a dry matter basis [19,20,22,24,26,27,29–31,36,37,39,41], and two articles reported details of the BSF larvae total carbohydrate content, on a dry matter basis [39,41]. One article reported additional details of the BSF larvae total lipid content of a fresh weight basis [37].

Rearing substrates 'Fruit and vegetables (uncharacterised)' reared for an unspecified number of days [39], 'Pomace, all-year mix (ratio of 5:5)' reared for 26 days and 'Pineapple' reared for 33.7 days [19] resulted in the lowest reported total lipid content of 2.2%, 4.9%

and 5.0% total lipid of BSF larvae dry matter, respectively. The highest BSF larvae total lipid content was reported for larvae reared for 14 days on 'Bread', resulting in 57.8% total lipid BSF larvae dry matter [24]. This was followed by 'Spent barley, brewer's yeast' reared for an unspecified number of days [22] and 'Fish *O. mykiss*, wheat (ratio of 5:1)' reared for 14 days [24] (49.0% and 46.7% total lipid of BSF larvae dry matter, respectively).

Of the two articles reporting on BSF larvae total carbohydrate content (a combined total of three rearing substrates), 'Fish waste (uncharacterised)' larvae reared for 12 days produced the highest reported result of 12.7% total carbohydrate BSF larvae dry matter [40]. This [40] and 'Fruit and vegetable (uncharacterised)' reared for an unspecified number of days [39] (12.3% and 8.4% total carbohydrate of BSF larvae dry matter, respectively).

The one article reporting BSF larvae nutritional composition on a fresh weight basis showed the highest protein in BSF larvae reared on 'Fruit and vegetable mix (ratio of 1:1)' for 36.7 days (17.6% of fresh weight) and highest lipid content in BSF larvae reared on 'Fruit—apple, pear, orange (ratio of 3.3:3.3:3.3)' for 52 days (21.0% of fresh weight) [37].

3.7. Essential Amino Acid Composition of Black Soldier Fly Larvae Reared on Food Waste

Seven articles included in this review provided details on the essential amino acid composition of BSF larvae reared on different substrates. Of these, six articles reported BSF larvae essential amino acid composition on a dry matter basis [26,32,33,35,36,38], with one article reporting on a wet weight basis (Table 4) [41].

Table 4. Essential Amino Acid Profile of Black Soldier Fly Larvae Reared on Food Waste.

Author	Rearing Substrate (RS) (Mixture Ratio)	Black Soldier Fly Larvae Essential Amino Acids								
		Histidine (%)	Isoleucine (%)	Leucine (%)	Lysine (%)	Methionine (%)	Phenylalanine (%)	Threonine (%)	Tryptophan (%)	Valine (%)
Liland [26]	RS 1: Wheat	2.8	3.9	6.4	6.2	1.7	4.0	3.9	-	5.8
	RS 2: Wheat, brown algae *A. nodosum* (9:1)	2.6	4.0	6.6	6.2	1.6	3.9	4.0	-	6.0
	RS 3: Wheat, brown algae *A. nodosum* (8:2)	2.7	4.0	6.7	5.9	1.7	4.3	4.1	-	6.0
	RS 4: Wheat, brown algae *A. nodosum* (7:3)	2.4	3.9	6.6	6.0	1.6	3.8	4.0	-	6.0
	RS 5: Wheat, brown algae *A. nodosum* (6:4)	2.5	4.0	6.6	5.5	1.5	3.7	3.9	-	5.7
	RS 6: Wheat, brown algae *A. nodosum* (5:5)	2.4	4.0	6.7	5.6	1.4	3.4	4.0	-	5.7
	RS 7: Wheat, brown algae *A. nodosum* (4:6)	2.5	4.1	6.9	5.5	1.5	3.6	4.1	-	5.6
	RS 8: Wheat, brown algae *A. nodosum* (3:7)	2.3	3.8	6.3	5.6	1.3	3.0	3.7	-	5.4
	RS 9: Wheat, brown algae *A. nodosum* (2:8)	2.5	3.8	6.3	5.4	1.4	3.2	3.9	-	5.4
	RS 10: Wheat, brown algae *A. nodosum* (1:9)	2.5	3.7	6.2	5.5	1.3	3.0	3.9	-	5.5
	RS 11: Brown algae *A. nodosum*	2.7	3.8	6.2	5.6	1.4	3.2	3.9	-	5.5
Shumo [+] [32]	RS 1: Kitchen waste—potato peelings, carrot, rice, bread debris (ratio unspecified)	0.3	0.3	0.3	0.5	0.8	0.5	-	-	0.1
	RS 2: Brewery by-product spent grain	0.5	0.2	0.4	0.5	0.7	0.2	-	-	0.9
Spranghers [^] [33]	RS 1: Restaurant waste—potato, rice, pasta, vegetables (ratio unspecified)	1.4	1.9	3.1	2.3	0.7	1.6	1.6	0.5	2.8
Tschirner [35]	RS 1: Carbohydrate—wheat middlings	3.3	4.2	6.6	5.9	1.6	3.6	3.9	-	5.7
	RS 2: Protein—dried distillers' grains with soluble	-	-	-	-	-	-	-	-	-
	RS 3: Fibre—sugar beet	-	-	-	-	-	-	-	-	-
Cappellozza [+] [36]	RS 1: Fruit and vegetable mix—zucchini, apple, potato, green beans, carrot, pepper, orange, celery, kiwi, plum, eggplant (unspecified ratio)	1.2	1.6	2.7	2.0	1.8	1.9	2.2	0.4	2.5
Lalander [^] [38]	RS 1: Food waste (uncharacterised)	2.9	4.1	6.8	8.3	1.8	4.0	3.9	1.4	5.8
	RS 2: Fruit and vegetable mix—lettuce, apple, potato (5:3:2)	2.6	4.3	6.7	5.1	1.5	3.5	3.5	1.4	6.0
Surendra [WW] [41]	RS 1: Food waste (uncharacterised)	1.7	1.5	2.3	2.2	0.9	1.5	1.5	-	2.4

Results presented as a percentage of the BSF larvae total protein content unless indicated. Liland et al. [26] presented as percentage of total sum of amino acids. [^] indicative of original article presenting data as g/kg of the BSF larvae total protein content, [+] indicative of original article presenting data as mg/g of the total BSF larvae total protein. Dashes used to indicate where data is unreported in the original article. [WW] wet weight.

Histidine—Histidine comprised 2.5% of BSF larvae total protein content when reported on a dry matter basis. Rearing BSF larvae on 'Kitchen waste—potato peelings, carrot, rice, bread debris (ratio unspecified)' for an unspecified number of days, resulted in the lowest histidine content of 0.3% of the total larval protein content [32]. The highest BSF larvae histidine content was reported for BSF larvae reared for 15 days on 'Carbohydrate—wheat middlings' with 3.3% of BSF larvae total protein content [35].

Isoleucine—Isoleucine comprised 3.8% of BSF larvae total protein content when reported on a dry matter basis. Rearing substrate 'Brewery by-product spent grain' reared for an unspecified number of days, resulted in the lowest reported isoleucine content of 0.2% of BSF larvae total protein content [32]. The highest BSF larvae isoleucine content was reported for BSF larvae reared for 43–47 days on 'Fruit and vegetable mix—lettuce, apple, potato (5:3:2)' with 4.3% of BSF larvae total protein content [38].

Leucine—Leucine was the most abundant essential amino acid and comprised 6.3% of BSF larvae total protein content when reported on a dry matter basis. Rearing substrate 'Kitchen waste—potato peelings, carrot, rice, bread debris (ratio unspecified)' reared for an unspecified number of days, resulted in the lowest reported leucine content of 0.3% of BSF larvae total protein content [32]. The highest BSF larvae leucine content was reported for BSF larvae reared for eight days on 'Wheat, brown algae *A. nodosum* (4:6)' with 6.9% of BSF larvae total protein content [26].

Lysine—Lysine comprised 5.6% of BSF larvae total protein content when reported on a dry matter basis. Rearing substrates 'Kitchen waste—potato peelings, carrot, rice, bread debris (ratio unspecified)' and 'Brewery by-product spent grain' reared for an unspecified number of days, both resulting in the lowest reported lysine content of 0.5% of BSF larvae total protein content [32]. The highest BSF larvae lysine content was reported for BSF larvae reared for 19 days on 'Food waste (uncharacterised)' with 8.3% of BSF larvae total protein content [38].

Methionine—Methionine comprised 1.5% of BSF larvae total protein content when reported on a dry matter basis. Rearing substrates 'Brewery by-product spent grain' reared for an unspecified number of days [32] and 'Restaurant waste—potato, rice, pasta, vegetables (ratio unspecified)' reared for 19 days [33] both resulting in the lowest reported methionine content of 0.7% of BSF larvae total protein content. The highest BSF larvae methionine content was reported for BSF larvae reared for 45 days on 'Fruit and vegetable mix—zucchini, apple, potato, green beans, carrot, pepper, orange, celery, kiwi, plum, eggplant (unspecified ratio)' [36] and 'Food waste (uncharacterised)' reared for 19 days, both resulting in 1.8% methionine of BSF larvae total protein content [38].

Phenylalanine—Phenylalanine comprised of 3.5% of BSF larvae total protein content when reported on a dry matter basis. Rearing substrate 'Brewery by-product spent grain' reared for an unspecified number of days, resulted in the lowest reported phenylalanine content of 0.2% of BSF larvae total protein content [32]. The highest BSF larvae phenylalanine content was reported for BSF larvae reared for eight days on 'Wheat, brown algae *A. nodosum* (8:2)' with 4.3% of BSF larvae total protein content [26].

Threonine—Threonine comprised 3.8% of BSF larvae total protein content when reported on a dry matter basis. Rearing substrate 'Restaurant waste—potato, rice, pasta, vegetables (ratio unspecified)' reared for 19 days, resulted in the lowest reported threonine content of 1.6% of BSF larvae total protein content [33]. The highest BSF larvae threonine content was reported for BSF larvae reared for eight days on 'Wheat, brown algae *A. nodosum* (8:2)' and 'Wheat, brown algae *A. nodosum* (4:6)' both resulting in 4.1% threonine of BSF larvae total protein content [26].

Tryptophan—Tryptophan was the least abundant essential amino acid and comprised 1.1% of BSF larvae total protein content when reported on a dry matter basis. Rearing substrate 'Fruit and vegetable mix—zucchini, apple, potato, green beans, carrot, pepper, orange, celery, kiwi, plum, eggplant (unspecified ratio)' reared for 45 days, resulted in the lowest reported tryptophan content of 0.4% of BSF larvae total protein content [36]. The highest BSF larvae tryptophan content was reported for BSF larvae reared for 19 days on 'Food waste (uncharacterised)' and BSF larvae reared for 42–47 days on 'Fruit and vegetable mix—lettuce, apple, potato (5:3:2)' both resulting in 1.4% tryptophan of BSF larvae total protein content [38].

Valine—Valine comprised 5.5% of BSF larvae total protein content when reported on a dry matter basis. Rearing substrate 'Kitchen waste—potato peelings, carrot, rice, bread debris (ratio unspecified)' reared for an unspecified number of days, resulted in the lowest valine content of 0.1% of BSF larvae total protein content [32]. The highest BSF larvae valine content was reported for BSF larvae reared for eight days on 'Wheat, brown algae *Ascophyllum nodosum* (9:1, 8:2 and 7:3)' all resulting in 6.0% valine of BSF larvae total protein content [26].

The one article reporting BSF larvae essential amino acid composition on a wet weight basis showed the least and most abundant essential amino acid to be methionine (0.9% of BSF larvae total protein content) and valine (2.4% of BSF larvae total protein content), respectively, when reared on 'Food waste (uncharacterised)' for an unspecified number of days [41].

3.8. Non-Essential Amino Acid Composition of Black Soldier Fly Larvae Reared on Food Waste

Seven articles included in this review provided details on the non-essential amino acid composition of BSF larvae reared on different substrates. Of these, six articles reported BSF larvae non-essential amino acid composition on a dry matter basis [26,32,33,35,36,38], with one article reporting on a wet weight basis (Table 5) [41].

Table 5. Non-Essential Amino Acid Profile of Black Soldier Fly Larvae Reared on Food Waste.

Author	Rearing Substrate (RS) (Mixture Ratio)	Black Soldier Fly Larvae Non-Essential Amino Acids										
		Alanine (%)	Arginine (%)	Aspartate (%)	Cysteine (%)	Glutamate (%)	Glutamic acid (%)	Glutamine (%)	Glycine (%)	Proline (%)	Serine (%)	Tyrosine (%)
Liland [26]	RS 1: Wheat	6.2	4.6	9.4	-	-	10.3	-	4.6	5.3	4.0	5.7
	RS 2: Wheat, brown algae *A. nodosum* (9:1)	6.6	4.5	9.2	-	-	11.8	-	5.0	5.8	4.3	5.6
	RS 3: Wheat, brown algae *A. nodosum* (8:2)	6.5	4.9	8.6	-	-	11.3	-	5.2	5.8	4.4	5.6
	RS 4: Wheat, brown algae *A. nodosum* (7:3)	6.6	4.5	8.5	-	-	11.9	-	5.1	5.8	4.4	5.1
	RS 5: Wheat, brown algae *A. nodosum* (6:4)	6.5	5.0	8.1	-	-	11.8	-	5.0	5.9	4.5	4.9
	RS 6: Wheat, brown algae *A. nodosum* (5:5)	6.8	4.6	8.3	-	-	12.3	-	4.8	5.9	4.6	4.3
	RS 7: Wheat, brown algae *A. nodosum* (4:6)	6.9	5.3	8.2	-	-	12.8	-	5.3	6.1	4.9	4.4
	RS 8: Wheat, brown algae *A. nodosum* (3:7)	6.6	4.5	7.9	-	-	11.7	-	4.7	5.4	4.3	3.6
	RS 9: Wheat, brown algae *A. nodosum* (2:8)	6.5	5.0	8.0	-	-	11.9	-	5.0	5.5	4.4	4.1
	RS 10: Wheat, brown algae *A. nodosum* (1:9)	6.3	4.7	8.0	-	-	11.6	-	5.0	5.0	4.4	4.0
	RS 11: Brown algae *A. nodosum*	6.4	6.5	8.3	-	-	11.9	-	4.3	5.2	4.3	4.2
Shumo [+] [32]	RS 1: Kitchen waste—potato peelings, carrot, rice, bread debris (ratio unspecified)	-	0.5	-	-	-	0.6	0.8	-	0.5	-	0.5
	RS 2: Brewery by-product spent grain	-	0.3	-	-	-	0.3	0.0	-	0.2	-	0.3
Spranghers ^ [33]	RS 1: Restaurant waste—potato, rice, pasta, vegetable (ratio unspecified)	2.8	2.0	3.7	0.2	-	4.6	-	2.5	2.5	1.6	-
Tschirner [35]	RS 1: Carbohydrate—wheat middlings	7.8	4.8	8.2	0.9	-	11.8	-	5.6	6.2	4.3	5.1
	RS 2: Protein—dried distillers' grains with soluble	-	-	-	-	-	-	-	-	-	-	-
	RS 3: Fibre—sugar beet	-	-	-	-	-	-	-	-	-	-	-
Cappellozza [+] [36]	RS 1: Fruit and vegetable mix—zucchini, apple, potato, green beans, carrot, pepper, orange, celery, kiwi, plum, eggplant (unspecified ratio)	3.9	4.8	3.8	0.5	-	6.5	-	2.3	1.5	1.8	2.0
Lalander ^ [38]	RS 1: Food waste (uncharacterised)	5.9	4.9	9.1	0.5	9.8	-	-	5.3	5.1	4.1	6.0
	RS 2: Fruit and vegetable mix—lettuce, apple, potato (5:3:2)	5.5	4.5	8.1	0.5	9.5	-	-	5.2	5.3	3.9	5.5
Surendra [WW] [41]	RS 1: Food waste (uncharacterised)	2.7	2.2	-	1.1	-	-	-	2.5	2.1	1.5	2.4

Results presented as a percentage of the BSF larvae total protein content unless indicated. Liland et al. [26] presented as percentage of total sum of amino acids. ^ indicative of original article presenting data as g/kg of the BSF larvae total protein content, [+] indicative of original article presenting data as mg/g of the total BSF larvae total protein. Dashes used to indicate where data is unreported in the original article. [WW] wet weight.

Alanine—Alanine comprised 6.2% of BSF larvae total protein content when reported on a dry matter basis. Rearing substrate 'Restaurant waste—potato, rice, pasta, vegetable (ratio unspecified)' reared for 19 days, resulted in the lowest reported alanine content of 2.8% of BSF larvae total protein content [33]. The highest BSF larvae alanine content was reported for BSF larvae reared for 15 days on 'Carbohydrate—wheat middlings with 7.8% of BSF larvae total protein content [35].

Arginine—Arginine comprised 4.3% of BSF larvae total protein content when reported on a dry matter basis. Rearing substrate 'Brewery by-product spent grain' reared for an unspecified number of days, resulted in the lowest reported arginine content of 0.3% of BSF larvae total protein content [32]. The highest BSF larvae arginine content was reported for BSF larvae reared for eight days on 'Brown algae *A. nodosum*' with 6.5% of BSF larvae total protein content [26].

Aspartate—Aspartate comprised 8.1% of BSF larvae total protein content when reported on a dry matter basis. Rearing substrate 'Restaurant waste—potato, rice, pasta, vegetable (ratio unspecified)' reared for 19 days, resulted in the lowest reported aspartate content of 3.7% of BSF larvae total protein content [33]. The highest BSF larvae aspartate content was reported for BSF larvae reared for eight days on 'Wheat' with 9.4% of BSF larvae total protein content [26].

Cysteine—Cysteine comprised 0.5% of BSF larvae total protein content when reported on a dry matter basis. Rearing substrate 'Fruit and vegetable mix—zucchini, apple, potato, green beans, carrot, pepper, orange, celery, kiwi, plum, eggplant (unspecified ratio)' reared for 45 days, resulted in the lowest reported cysteine content of 0.05% of BSF larvae total protein content [36]. The highest BSF larvae cysteine content was reported for BSF larvae reared for 15 days on 'Carbohydrate—wheat middlings with 0.9% of BSF larvae total protein content [35].

Glutamate—Of the one article (two rearing substrates) reporting BSF larvae glutamate composition, there was an average of 9.7% glutamate of BSF larvae total protein content when reported on a dry matter basis. Rearing substrate 'Fruit and vegetable mix—lettuce, apple, potato (5:3:2)' reared for 42–47 days, resulted in the lowest reported glutamate content of 9.5% of BSF larvae total protein content [38]. The highest BSF larvae glutamate content was reported for BSF larvae reared for 19 days on 'Food waste (uncharacterised)' with 9.8% of BSF larvae total protein content [38].

Glutamic acid—Glutamic acid was the most abundant non-essential amino acid and comprised 9.8% of BSF larvae total protein content when reported on a dry matter basis. Rearing substrate 'Brewery by-product spent grain' reared for an unspecified number of days, resulted in the lowest reported glutamic acid content of 0.3% of BSF larvae total protein content [32]. The highest BSF larvae glutamic acid content was reported for BSF larvae reared for eight days on 'Wheat, brown algae *A. nodosum* (4:6)' with 12.8% of BSF larvae total protein content [26].

Glutamine—One article (two rearing substrates) reporting BSF larvae glutamine composition, there was an average of 0.4% glutamine of BSF larvae total protein content when reported on a dry matter basis and was the least abundant non-essential amino acid. Rearing substrate 'Brewery by-product spent grain' for an unspecified number of days, resulted in the lowest reported glutamine content of 0.0% of BSF larvae total protein content. The highest BSF larvae glutamine content was reported for BSF larvae reared for an unspecified number of days on 'Kitchen waste—potato peelings, carrot, rice, bread debris (ratio unspecified)' with 0.8% of BSF larvae total protein content [32].

Glycine—Glycine comprised 4.8% of BSF larvae total protein content when reported on a dry matter basis. Rearing substrate 'Fruit and vegetable mix—zucchini, apple, potato, green beans, carrot, pepper, orange, celery, kiwi, plum, eggplant (unspecified ratio)' reared for 45 days, resulted in the lowest reported glycine content of 2.3% of BSF larvae total protein content [36]. The highest BSF larvae glycine content was reported for BSF larvae reared for 15 days on 'Carbohydrate—wheat middlings with 5.6% of BSF larvae total protein content [35].

Proline—Proline comprised 4.6% of BSF larvae total protein content when reported on a dry matter basis. Rearing substrate 'Brewery by-product spent grain' reared for an unspecified number of days, resulted in the lowest reported proline content of 0.2% of BSF larvae total protein content [32]. The highest BSF larvae proline content was reported for BSF larvae reared for 15 days on 'Carbohydrate—wheat middlings with 6.2% of BSF larvae total protein content [35].

Serine—Serine comprised 4.1% of BSF larvae total protein content when reported on a dry matter basis. Rearing substrate 'Restaurant waste—potato, rice, pasta, vegetable (ratio unspecified)' reared for 19 days, resulted in the lowest reported serine content of 1.6% of BSF larvae total protein content [33]. The highest BSF larvae serine content was reported for BSF larvae reared for eight days on 'Wheat, brown algae *A. nodosum* (4:6)' with 4.9% of BSF larvae total protein content [26].

Tyrosine—Tyrosine comprised 4.1% of BSF larvae total protein content when reported on a dry matter basis. Rearing substrate 'Brewery by-product spent grain' reared for an unspecified number of days, resulted in the lowest reported tyrosine content of 0.3% of BSF larvae total protein content [32]. The highest BSF larvae tyrosine content was reported for BSF larvae reared for 19 days on 'Food waste (uncharacterised)' with 6.0% of BSF larvae total protein content [38].

The one article reporting BSF larvae non-essential amino acid composition on a wet weight basis showed the least and most abundant non-essential amino acid to be cysteine (1.1% of BSF larvae total protein content) and alanine (2.7% of BSF larvae total protein content), respectively, when reared on 'Food waste (uncharacterised)' for an unspecified number of days [41].

4. Risk of Bias Assessment

Risk of Bias (ROB) assessment was performed using the Systematic Review Centre for Laboratory Animal Experimentation (SYRCLE) risk of bias tool [18], with the exclusion of items 5 and 7 (Supplementary Table S2). These items were removed from assessment following the instruction of the (SYRCLE) risk of bias tool to adapt the list to the specific needs of the review [18], with 'blinding of the caregiver' and 'blinding of the assessors' deemed by the authors as unlikely to influence the potential bias of the articles reviewed. This ROB assessment performed in this review highlights a potential widespread nature of poor method reporting and lack of intent to reduce the risk of bias in BSF larvae rearing investigations. All articles included in this review were missing information to various degrees, with all showing potential for selection bias due to the absence of methodological information reporting on allocation concealment or sequencing generation and as such, being at risk of a significant level of bias.

5. Discussion

BSF larvae offer a feasible and cost-effective solution to two growing global challenges: food waste management and the rising global demand for sustainable protein. For this decomposer insect to be utilised in the treatment of food waste, and then to be effectively implemented and accepted into the food supply, it is essential to further our knowledge regarding the influence of various food waste streams on the nutritional composition of the BSF larvae.

Dietary protein is an essential macronutrient in BSF larval development and is necessary for supporting adequate protein and lipid accumulation in the fat cells of the BSF [46]. The results of this review indicate the total protein content of the BSF larvae can be substantially influenced by the substrate protein content of which it is reared on, with BSF larvae reared for up to 28.7 days on various combinations of low protein 'Fruit and vegetable' substrates producing the least abundant source of BSF larval protein (12.9% total protein) [19,39]. In comparison, BSF larvae reared for two days on a high protein substrate 'Fish waste *S. aurita*' produced the most abundant source of BSF larval protein (78.8% total protein) [40].

Interestingly, BSF larvae reared on 'Fish waste *S. aurita*' for longer than two days displayed a steady decline in larval total protein content, suggesting that two days of rearing on this high protein food waste substrate would be the optimal condition to generate high protein BSF larvae [40]. Variation in the protein composition of the BSF larvae throughout their lifespan has been supported by others using non-food waste rearing substrates (Chicken feed), in which total protein has been reported to range between 34% and 42% during larval stages and between 31% and 46% [47–50]. The articles collated in this review began experimental feeding procedures at various stages in the BSF larvae life cycle, from placing eggs directly onto the rearing diets [22,36] to waiting until the BSF larvae were aged up to 14 days before introducing them to food waste substrates [25]. Furthermore, some studies reported different harvest stages, including the sight of first prepupae [30], when 40% of BSF larvae had reached prepupae [36,37] and when 100% of BSF larvae had reached prepupae [38] (Supplementary Table S3). As such, the composition of total protein in the BSF larvae reported across different studies may be compounded by factors such as age and harvest stage, rather than a representation of the type of food waste provisions.

The articles included in this review also reported widespread difference in the quantity of feed rations provided to the BSF larvae, with a range from 12.2 mg per larvae [23] to 1530 mg per larvae [24]. Insufficient feed rations have been shown to impair biomass production and influence BSF larvae protein content [48]. Whilst the optimal feeding ration of both chicken feed and fecal sludge has been determined for optimising BSF larvae biomass and nutritional content, the literature on the ideal food waste provision is limited [51]. This is likely due to the variations and inconsistencies in food waste products from a macronutrient perspective. It is plausible that the BSF larvae in the articles included in this review, may have been provided with less than adequate feed rations and as a result their protein composition was influenced by lack of sustenance as opposed to the macronutrient content of the rearing substrate provided.

In addition to rearing conditions influencing the composition of total protein in the BSF larvae, different data analyses methods used by different studies may have impacted on the ability to accurately compare the efficacy of differing food waste substrates. The total protein content of BSF larvae is most commonly determined from the total elemental nitrogen content using the common Kjeldahl method and the standard protein conversion factor 6.25 [52]. However, due to the additional non-protein nitrogen found in the insects' chitin, it is possible to over-estimate total protein content and as such a factor of 4.67 has been proposed as a more accurate representation of total protein content in insects [52]. When presenting the proximate composition of BSF larvae, Ewald et al. [24] included results of both conversion factors highlighting the differences between both factors. BSF larvae reared on 'Bread' for 14 days, with protein determination calculated with a conversion factor of 6.25, indicated the substrate to result in a total protein content of 39.2% of BSF larvae dry matter [24]. Yet, when the same data were reanalysed using a conversion factor of 4.67, a total protein content of 29.8% of BSF larvae dry matter was indicated; a substantial difference of 9.4% in the total protein content reported in BSF larvae [24]. Spranghers et al. [33] also included chitin corrected values when observing the influence of 'Restaurant waste—potato, rice, pasta, vegetable (ratio unspecified)' on the nutritional composition of the BSF, finding a similar decrease of 9.0% in BSF larvae total protein content when compared to data not taking chitin into consideration [33]. It is possible to include chitin corrected value, as this is obtained by analysing the nitrogen content of the chitin fraction and subtracting this from total nitrogen content, yet the reporting of the conversion factors was absent in many articles included in this review [20,21,23,36,37,39], as was the reporting of chitin correction adjustments [19–22,24–28,30–32,34,35,37–41], which may hinder the ability to draw comparisons between the total protein content of BSF larvae reared on various food waste substrates included in this review.

The BSF larvae's amino acid profile has been shown to be suitable for use as pet food [47] and animal feed [41]. To date, only a few studies have examined the relationship between the substrate amino acid composition and that of the BSF larvae [25,32]. When observing the impact of food waste substrates on the amino acid profiles of BSF larvae, of the two articles included in this review, glutamic acid was reported to be the most abundant non-essential amino acid (when reared on 'Wheat, brown algae *A. nodosum* (4:6)' for eight days) [26] and leucine the most abundant essential amino acid (when reared on 'Wheat, brown algae *A. nodosum* (4:6)' for eight days) [26]. This was consistent with the glutamic acid and leucine content also being reported as most prevalent in the food waste rearing substrates provided to BSF larvae [26,33]. Aside from both glutamic acid and leucine, there was a great variation in the amino acid content of substrates used in different studies (Supplementary Table S1.). Despite this, the BSF larvae only exhibited minimal differences in amino acid content (±20%) within each study [33,38]. This would indicate that the amino acid content of the BSF larvae has a limited opportunity for manipulation when reared on food waste products regardless of the amino acid content of the rearing substrate provided. With only two articles providing amino acid content of both BSF larvae and the rearing substrate, there is limited information available to draw a solid conclusion on the role of rearing substrates in the influence of the amino acid content on BSF larvae. This makes further studies essential to developing a clearer understand of this relationship.

The choice of processing methods has been shown to influence the amino acid content of the BSF larvae, including the culling and drying method used in preparing the BSF larvae [53]. Articles included in this review reported various drying techniques including freezing [23,34] and freeze-drying in liquid nitrogen [29]. Huang et al. [53] reported that conventional drying (60.0 °C) of BSF larvae produced a higher digestible indispensable amino acid score when compared with microwave drying of BSF larvae. Others have reported that culling BSF larvae by a method of freezing resulted in a reduction in amino acids cysteine and lysine content [54]. Both Liland et al. [26] and Spranghers et al. [33] reporting freezing as their method of culling BSF larvae, there is an increased likelihood of the amino acid content values of the BSF larvae being inaccurately reported.

In addition to rearing substrate, various abiotic factors can affect the development of BSF larvae and may explain the variability of the nutritional content of BSF larvae in the studies included in this review. Abiotic factors that may influence BSF larvae development include larval density [55], larval handling [42], substrate moisture content or pH level [56], however, these factors were not extracted from articles and taken into consideration in this current review.

6. Conclusions and Future Directions

The results of this review on the influence of food waste products on BSF larvae protein content infer that the total protein content of food waste products used as a rearing substrate is likely to result in producing BSF larvae with a similar total protein and amino acid content. However, due to the variation in methodologies applied within each study and absence of BSF larvae nutritional composition at the commencement of the studies, there is a reduced confidence in the extent to which these various food waste substrates may have influenced the total protein content of BSF larvae. The standardisation of methodologies in BSF larvae resource conversion studies have been proposed by others and adherence to a standard methodology may increase confidence in future studies [44].

The transformation and nutrient recovery prospects of using BSF larvae as a food waste management system and as a sustainable protein source are promising. However, further research is required regarding the influence of various food waste streams on the protein composition outcomes of the BSF larvae, as well as a greater understanding of the potential influence of anti-nutritive elements on the nutritional profiles of BSF larvae. Further research exploring these factors will improve the successful introduction of BSF larvae as a novel feed and/or food.

Supplementary Materials: The following are available online at https://www.mdpi.com/article/10.3390/insects12070608/s1, Supplementary Table S1, Amino Acid Profile of Rearing Substrates. Supplementary Table S2, Risk of bias assessment. Supplementary Table S3, Life history traits of Black Soldier Fly Larvae Reared on Food Waste.

Author Contributions: All authors contributed to the development of the research protocol (I.H., L.P.N., H.G. and J.D.). Titles and abstracts collected from the search were screened by two investigators (I.H. and J.D.). In circumstances where it was not clear whether an article met the eligibility criteria, full-text articles were obtained, screened and resolved by discussion and consensus between three investigators (I.H., L.P.N., H.G. and J.D.). All authors provided intellectual input to the final manuscript (I.H., L.P.N., H.G. and J.D.). All authors have read and agreed to the published version of the manuscript.

Funding: This research received no external funding.

Institutional Review Board Statement: Not applicable.

Data Availability Statement: The data presented in this study are available in The Influence of Food Waste Rearing Substrates on Black Soldier Fly Larvae Protein Composition: A Systematic Review.

Conflicts of Interest: The authors declare no conflict of interest.

References

1. United Nations Department of Economic and Social Affairs, Population Division. World Population Prospects 2019: Highlights. Available online: Population.un.org/wpp/Publications/Files/WPP2019_Highlights.pdf (accessed on 28 September 2020).
2. Food and Agriculture Organization of the United Nations, Department of Economic and Social Affairs, Population Division. Feeding the World in 2050. Available online: http://www.fao.org/tempref/docrep/fao/meeting/018/k6021e.pdf (accessed on 28 September 2020).
3. Food and Agriculture Organization. The State of Food and Agriculture 2019, Moving forward on Food Loss and Waste Reduction. Available online: http://www.fao.org/3/ca6030en/ca6030en.pdf (accessed on 12 April 2021).
4. Gustavsson, J.; Cederberg, C.; Sonesson, U. Global Food Losses and Food Waste—Extent, Causes and Prevention. Food and Agriculture Organization. Available online: http://www.fao.org/3/mb060e/mb060e.pdf (accessed on 7 October 2020).
5. Environment Victoria. The Problem with Landfill. Available online: https://environmentvictoria.org.au/resource/problem-landfill/ (accessed on 7 October 2020).
6. Scialabba, N.; Jan, O.; Tostivint, C.; Turbé, A.; O'Connor, C.; Lavelle, P.; Flammini, A.; Hoogeveen, J.; Iweins, M.; Tubiello, F.; et al. Food Wastage Footprint: Impacts on Natural Resources. Summary Report. Food and Agriculture Organization of the United Nations 2013. Available online: http://www.fao.org/3/i3347e/i3347e.pdf (accessed on 7 October 2020).
7. Liu, C.; Wang, C.; Yao, H. Comprehensive resource utilization of waste using the Black Soldier Fly (*Hermetia illucens* (L.)) (*Diptera: Stratiomyidae*). *Animals* **2019**, *9*, 349. [CrossRef]
8. Diener, S.; Studt Solano, N.M.; Roa Gutierrez, F.; Zurbruegg, C.; Tockner, K. Biological treatment of municipal organic waste using Black Soldier Fly larvae. *Waste Biomass Valorization* **2011**, *2*, 357–363. [CrossRef]
9. Mohd-Noor, S.N.; Wong, C.Y.; Lim, J.W.; Mah-Hussin, M.; Uemura, Y.; Lam, M.K.; Ramli, A.; Bashir, M.J.K.; Tham, L. Optimization of self-fermented period of waste coconut endosperm destined to feed Black Soldier Fly larvae in enhancing the lipid and protein yields. *Renew. Energy* **2017**, *111*, 646–654. [CrossRef]
10. Parra Paz, A.S.; Carrejo, N.S.; Gómez Rodríguez, C.H. Effects of larval density and feeding rates on the bioconversion of vegetable waste using Black Soldier Fly larvae *Hermetia illucens* (L.), (*Diptera: Stratiomyidae*). *Waste Biomass Valorization* **2015**, *6*, 1059–1065. [CrossRef]
11. Zheng, L.Y.; Hou, Y.F.; Li, W.; Yang, S.; Li, Q.; Yu, Z.N. Biodiesel production from rice straw and restaurant waste employing Black Soldier Fly assisted by microbes. *Energy* **2012**, *47*, 225–229. [CrossRef]
12. Wong, C.; Rosli, S.; Uemura, Y.; Ho Rosli, S.S.; Uemura, Y.; Ho, Y.C.; Leejeerajumnean, A.; Kiatkittipong, W.; Cheng, C.K.; Lam, M.K.; et al. Potential protein and biodiesel sources from Black Soldier Fly larvae: Insights of larval harvesting instar and fermented feeding medium. *Energies* **2019**, *12*, 1570. [CrossRef]
13. Belghit, I.; Liland, N.S.; Gjesdal, P.; Biancarosa, I.; Menchetti, E.; Li, Y.; Waagbo, R.; Krogdahl, A.; Lock, E.J. Black Soldier Fly larvae meal can replace fish meal in diets of sea-water phase Atlantic salmon (*Salmo salar*). *Aquaculture* **2019**, *503*, 609–619. [CrossRef]
14. Biasato, I.; Meneguz, M.; Bressan, E.; Dama, A.; Gasco, L.; Renna, M.; Dabbou, S.; Capucchio, M.T.; Schiavone, A.; Gai, F.; et al. Partially defatted Black Soldier Fly larva meal inclusion in piglet diets: Effects on the growth performance, nutrient digestibility, blood profile, gut morphology and histological features. *J. Anim. Sci. Biotechnol.* **2019**, *10*, 12. [CrossRef]
15. Van Huis, A.; Van Itterbeeck, J.; Klunder, H.; Mertens, E.; Halloran, A.; Muir, G.; Vantomme, P. *Edible Insects: Future Prospects for Food and Feed Security*; FAO: Rome, Italy, 2013; Volume 171, ISBN 978-92-5-107595-1.
16. Mitsuhashi, J. *Edible Insects of the World*; CRC Press: Boca Raton, FL, USA, 2016. [CrossRef]
17. Liberati, A.; Altman, D.G.; Tetzlaff, J.; Mulrow, C.; Gøtzsche, P.C.; Ioannidis, J.P.A.; Clarke, M.; Devereaux, P.J.; Kleijnen, J.; Moher, D. The PRISMA statement for reporting systematic reviews and meta-analyses of studies that evaluate healthcare interventions: Explanation and elaboration. *BMJ* **2009**, *339*, b2700. [CrossRef]
18. Hooijmans, C.R.; Rovers, M.M.; de Vries, R.B.M.; Leenaars, M.; Ritskes-Hoitinga, M.; Langendam, M.W. SYRCLE's risk of bias tool for animal studies. *BMC Med. Res. Methodol.* **2014**, *14*, 43. [CrossRef]
19. Barbi, S.; Macavei, L.I.; Fuso, A.; Luparelli, A.V.; Caligiani, A.; Ferrari, A.M.; Maistrello, L.; Montorsi, M. Valorization of seasonal agri-food leftovers through insects. *Sci. Total Environ.* **2020**, *709*, 136209. [CrossRef] [PubMed]

20. Barragán-Fonseca, K.; Pineda-Mejia, J.; Dicke, M.; Van Loon, J.J.A. Performance of the Black Soldier Fly (*Diptera: Stratiomyidae*) on Vegetable Residue-Based Diets Formulated Based on Protein and Carbohydrate Contents. *J. Econ. Entomol.* **2018**, *111*, 2676–2683. [CrossRef] [PubMed]

21. Bava, L.; Gislon, G.; Zucali, M.; Colombini, S.; Jucker, C.; Lupi, D.; Savoldelli, S. Rearing of *Hermetia Illucens* on different organic by-products: Influence on growth, waste reduction, and environmental impact. *Animals* **2019**, *9*, 289. [CrossRef] [PubMed]

22. Chia, S.Y.; Tanga, C.M.; Osuga, I.M.; Cheseto, X.; Ekesi, S.; Dicke, M.; Van Loon, J.J.A. Nutritional composition of Black Soldier Fly larvae feeding on agro-industrial by-products. *Entomol. Exp. Appl.* **2020**, *168*, 472–481. [CrossRef]

23. Danieli, P.P.; Amici, A.; Ronchi, B.; Lussiana, C.; Gasco, L. The Effects of Diet Formulation on the Yield, Proximate Composition, and Fatty Acid Profile of the Black Soldier Fly (*Hermetia illucens L.*) Prepupae Intended for Animal Feed. *Animals* **2019**, *9*, 178. [CrossRef]

24. Ewald, N.; Vidakovic, A.; Langeland, M.; Kiessling, A.; Sampels, S.; Lalander, C. Fatty acid composition of Black Soldier Fly larvae (*Hermetia illucens*)—Possibilities and limitations for modification through diet. *Waste Manag.* **2020**, *102*, 40–47. [CrossRef]

25. Gold, M.; Cassar, C.M.; Zurbrugg, C.; Kreuzer, M.; Boulos, S.; Diener, S.; Mathys, A. Biowaste treatment with Black Soldier Fly larvae: Increasing performance through the formulation of biowastes based on protein and carbohydrates. *Waste Manag.* **2020**, *102*, 319–329. [CrossRef]

26. Liland, N.S.; Biancarosa, I.; Araujo, P.; Biemans, D.; Bruckner, C.G.; Waagbo, R.; Torstensen, B.E.; Lock, E.J. Modulation of nutrient composition of Black Soldier Fly (*Hermetia illucens*) larvae by feeding seaweed-enriched media. *PLoS ONE* **2017**, *12*, e0183188. [CrossRef]

27. Liu, Z.; Minor, M.; Morel, P.C.H.; Najar-Rodriguez, A.J. Bioconversion of Three Organic Wastes by Black Soldier Fly (*Diptera: Stratiomyidae*) Larvae. *Environ. Entomol.* **2018**, *47*, 1609–1617. [CrossRef]

28. Lopes, I.G.; Lalander, C.; Vidotti, R.M.; Vinneras, B. Using *Hermetia illucens* larvae to process biowaste from aquaculture production. *J. Clean. Prod.* **2020**, *251*, 119753. [CrossRef]

29. Meneguz, M.; Dama, A.; Lussiana, C.; Renna, M.; Gasco, L.; Schiavone, A. Effect of rearing substrate on growth performance, waste reduction efficiency and chemical composition of Black Soldier Fly (*Hermetia illucens*) larvae. *J. Sci. Food Agric.* **2018**, *98*, 5776–5784. [CrossRef]

30. Oonincx, D.G.; Van Broekhoven, S.; Van Huis, A.; van Loon, J.J. Feed conversion, survival and development, and composition of four insect species on diets composed of food by-products. *PLoS ONE* **2015**, *10*, e0144601. [CrossRef] [PubMed]

31. Salomone, R.; Saija, G.; Mondello, G.; Giannetto, A.; Fasulo, S.; Savastano, D. Environmental impact of food waste bioconversion by insects: Application of Life Cycle Assessment to process using *Hermetia illucens*. *J. Clean. Prod.* **2017**, *140*, 890–905. [CrossRef]

32. Shumo, M.; Osuga, I.M.; Khamis, F.M.; Tanga, C.M.; Fiaboe, K.K.M.; Subramanian, S.; Ekesi, S.; Van Huis, A.; Borgemeister, C. The nutritive value of Black Soldier Fly larvae reared on common organic waste streams in Kenya. *Sci. Rep.* **2019**, *9*, 10110. [CrossRef]

33. Spranghers, T.; Ovyn, A.; De Smet, S.; De Clercq, P.; Ottoboni, M.; Ovyn, A.; Michiels, J.; Eeckhout, M.; Deboosere, S.; De Meulenaer, B. Nutritional composition of Black Soldier Fly (*Hermetia illucens*) prepupae reared on different organic waste substrates. *J. Sci. Food Agric.* **2016**, *97*, 2594–2600. [CrossRef] [PubMed]

34. Tinder, A.C.; Puckett, R.T.; Turner, N.D.; Cammack, J.A.; Tomberlin, J.K. Bioconversion of sorghum and cowpea by Black Soldier Fly (*Hermetia illucens* (L.)) larvae for alternative protein production. *J. Insects Food Feed* **2017**, *3*, 121–130. [CrossRef]

35. Tschirner, M.; Simon, A. Influence of different growing substrates and processing on the nutrient composition of Black Soldier Fly larvae destined for animal feed. *J. Insects Food Feed* **2015**, *1*, 249–259. [CrossRef]

36. Cappellozza, S.; Saviane, A.; Leonardi, M.G.; Savoldelli, S.; Carminati, D.; Rizzolo, A.; Cortellino, G.; Terova, G.; Bruno, D.; Tettamanti, G.; et al. A first attempt to produce proteins from insects by means of a circular economy. *Animals* **2019**, *9*, 278. [CrossRef]

37. Jucker, C.; Erba, D.; Leonardi, M.G.; Lupi, D.; Savoldelli, S. Assessment of vegetable and fruit substrates as potential rearing media for *Hermetia illucens* (*Diptera: Stratiomyidae*) larvae. *Environ. Entomol.* **2017**, *46*, 1415–1423. [CrossRef]

38. Lalander, C.; Diener, S.; Zurbrugg, C.; Vinneras, B. Effects of feedstock on larval development and process efficiency in waste treatment with Black Soldier Fly (*Hermetia illucens*). *J. Clean. Prod.* **2019**, *208*, 211–219. [CrossRef]

39. Nguyen, T.T.; Tomberlin, J.K.; Vanlaerhoven, S. Ability of Black Soldier Fly (*Diptera: Stratiomyidae*) larvae to recycle food waste. *Environ. Entomol.* **2015**, *44*, 406–410. [CrossRef] [PubMed]

40. Barroso, F.G.; Sanchez-Muros, M.J.; Rincon, M.A.; Rodriguez-Rodriguez, M.; Fabrikov, D.; Morote, E.; Guil-Guerrero, J.L. Production of n-3-rich insects by bioaccumulation of fishery waste. *J. Food Compos. Anal.* **2019**, *82*, 103237. [CrossRef]

41. Surendra, K.C.; Olivier, R.; Tomberlin, J.K.; Jha, R.; Khanal, S.K. Bioconversion of organic wastes into biodiesel and animal feed via insect farming. *Renew. Energy* **2016**, *98*, 197–202. [CrossRef]

42. Nguyen, T.T.; Tomberlin, J.K.; Vanlaerhoven, S. Influence of resources on *Hermetia illucens* (*Diptera: Stratiomyidae*) larval development. *J. Med. Entomol.* **2013**, *50*, 898–906. [CrossRef]

43. Barros, L.M.; Gutjahr AL, N.; Ferreira-Keppler, R.L.; Martins, R.T. Morphological description of the immature stages of *Hermetia illucens* (*Linnaeus, 1758*) (*Diptera: Stratiomyidae*). *Microsc. Res. Tech.* **2018**, *82*, 178–189. [CrossRef]

44. Bosch, G.; Oonincx DG, A.B.; Jordan, H.R.; Zhang, J.; Van Loon, J.J.A.; Van Huis, A.; Tomberlin, J.K. Standardisation of quantitative resource conversion studies with Black Soldier Fly larvae. *J. Insects Food Feed* **2020**, *6*, 95–109. [CrossRef]

45. Chia, S.Y.; Tanga, C.M.; Osuga, I.M.; Mohamed, S.A.; Khamis, F.M.; Salifu, D.; Sevgan, S.; Fiaboe, K.; Niassy, S.; van Loon, J.; et al. Effects of waste stream combinations from brewing industry on performance of Black Soldier Fly, *Hermetia illucens* (*Diptera: Stratiomyidae*). *PeerJ* **2018**, *6*, e5885. [CrossRef]

46. Pimentel, A.C.; Montali, A.; Bruno, D.; Tettamanti, G. Metabolic adjustment of the larval fat body in *Hermetia illucens* to dietary conditions. *J. Asia Pac. Entomol.* **2017**, *20*, 1307–1313. [CrossRef]

47. Liu, X.; Chen, X.; Wang, H.; Yang, Q.Q.; Rehman, K.U.; Li, W.; Cai, M.M.; Li, Q.; Mazza, L.; Zhang, J.B.; et al. Dynamic changes of nutrient composition throughout the entire life cycle of Black Soldier Fly. *PLoS ONE* **2017**, *12*, e0182601. [CrossRef]

48. Diener, S.; Zurbrugg, C.; Tockner, K. Conversion of organic material by Black Soldier Fly larvae: Establishing optimal feeding rates. *Waste Manag. Res.* **2009**, *27*, 603–610. [CrossRef] [PubMed]

49. St-Hilaire, S.; Sheppard, C.; Tomberlin, J.K.; Irving, S.; Newton, L.; McGuire, M.A.; Mosley, E.E.; Hardy, R.W.; Sealey, W. Fly prepupae as a feedstuff for rainbow trout, *Oncorhynchus mykiss*. *J. World Aquac. Soc.* **2007**, *38*, 59–67. [CrossRef]

50. Newton, L.; Sheppard, C.; Watson, D.W.; Burtle, G.; Dove, R. *Using the Black Soldier Fly, Hermetia Illucens, as a Value-Added Tool for the Management of Swine Manure*; Anim Poult Waste Manag Center, North Carolina State University: Raleigh, NC, USA, 2005; Volume 17.

51. Nyakeri, E.M.; Ayieko, M.A.; Amimo, F.A.; Salum, H.; Ogola, H.J.O. An optimal feeding strategy for Black Soldier Fly larvae biomass production and faecal sludge reduction. *J. Insects Food Feed* **2019**, *5*, 201–213. [CrossRef]

52. Janssen, R.H.; Vincken, J.-P.; van den Broek, L.A.M.; Fogliano, V.; Lakemond, C.M.M. Nitrogen-to-Protein conversion factors for three edible insects: *Tenebrio molitor*, *Alphitobius diaperinus*, and *Hermetia illucens*. *J. Agric. Food Chem.* **2017**, *65*, 2275–2278. [CrossRef]

53. Huang, C.; Feng, W.; Xiong, J.; Wang, T.; Wang, W.; Wang, C.; Yang, F. Impact of drying method on the nutritional value of the edible insect protein from Black Soldier Fly (*Hermetia illucens* L.) larvae: Amino acid composition, nutritional value evaluation, in vitro digestibility, and thermal properties. *Eur. Food Res. Technol.* **2019**, *245*, 11–21. [CrossRef]

54. Leni, G.; Caligiani, A.; Sforza, S. Killing method affects the browning and the quality of the protein fraction of Black Soldier Fly (*Hermetia illucens*) prepupae: A metabolomics and proteomic insight. *Food Res. Int.* **2019**, *115*, 116–125. [CrossRef] [PubMed]

55. Barragán-Fonseca, K.B.; Dicke, M.; Van Loon, J.J.A. Influence of larval density and dietary nutrient concentration on performance, body protein, and fat contents of Black Soldier Fly larvae (*Hermetia illucens*). *Entomol. Exp. Appl.* **2018**, *166*, 761–770. [CrossRef]

56. Meneguz, M.; Gasco, L.; Tomberlin, J.K. Impact of pH and feeding system on Black Soldier Fly (*Hermetia illucens*, L.; *Diptera: Stratiomyidae*) larval development. *PLoS ONE* **2018**, *13*, e0202591. [CrossRef] [PubMed]

Article

Growth Performance, Waste Reduction Efficiency and Nutritional Composition of Black Soldier Fly (*Hermetia illucens*) Larvae and Prepupae Reared on Coconut Endosperm and Soybean Curd Residue with or without Supplementation

Nichaphon Pliantiangtam [1], Pipatpong Chundang [2] and Attawit Kovitvadhi [2,*]

1   Animal Health and Biomedical Science Program, Faculty of Veterinary Medicine, Kasetsart University, 50 Ngamwongwan Rd., Bangkok 10900, Thailand; nichaphon.p@ku.th
2   Department of Physiology, Faculty of Veterinary Medicine, Kasetsart University, 50 Ngamwongwan Rd., Bangkok 10900, Thailand; pichandang@gmail.com
*   Correspondence: fvetawk@ku.ac.th

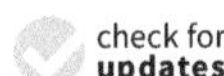

check for **updates**

Citation: Pliantiangtam, N.; Chundang, P.; Kovitvadhi, A. Growth Performance, Waste Reduction Efficiency and Nutritional Composition of Black Soldier Fly (*Hermetia illucens*) Larvae and Prepupae Reared on Coconut Endosperm and Soybean Curd Residue with or without Supplementation. *Insects* **2021**, *12*, 682. https://doi.org/10.3390/insects12080682

Academic Editors: Man P. Huynh, Kent S. Shelby and Thomas A. Coudron

Received: 7 July 2021
Accepted: 27 July 2021
Published: 29 July 2021

Publisher's Note: MDPI stays neutral with regard to jurisdictional claims in published maps and institutional affiliations.

Simple Summary: Black soldier fly (BSF, *Hermetia illucens*) larvae have a high potential to convert organic waste into high-value products. However, the growth performance, waste reduction efficiency, and chemical composition of BSF larvae are greatly influenced by the rearing substrate. This study focused on investigating the growth performance, waste reduction efficiency, and nutritional composition of BSF larvae reared on different ratios of coconut endosperm (C) and soybean curd residue (S), with or without supplementation, compared to standard diets (Gainesville: G and starter chicken diet: CK). The results showed that BSF larvae fed CK has the highest larval weight, followed by those fed coconut endosperm and soybean curd residue at a ratio of 20:80 (C20S80), and coconut endosperm and soybean curd residue at a ratio of 50:50 (C50S50) without supplementation. The greatest waste reduction efficiency was observed in the G, C50S50, and C20S80 groups without supplementation. The highest crude protein content in larvae was presented in the G and C20S80 groups followed by the CK and C50S50 groups. Therefore, equal proportions of C and S without supplementation is likely the best formulation for growth performance, waste reduction efficiency, and nutritional composition of BSF larvae when compared with standard diets.

Abstract: Black soldier fly (BSF, *Hermetia illucens*) larvae are considered as insects with a high potential to convert organic waste into high-value products. The objective of this study was to investigate the growth performance, waste reduction efficiency, and nutritional composition of BSF reared on different ratios of coconut endosperm (C) and soybean curd residue (S), with or without supplementation, compared to standard diets (Gainesville: G and starter chicken diet: CK). Seven-day-old larvae were randomly divided into eight experimental groups (G, CK, and three different ratios of C and S with or without supplementation) with three replicates with an equal weight of larvae. The supplement contained calcium, phosphorus, amino acids, and a mineral–vitamin premix which was formulated to correlate with CK. Each replicate was terminated, measured, and evaluated when 40% of larvae had reached prepupal stage. The highest larval weight gain was presented in BSF fed CK, followed by those fed coconut endosperm and soybean curd residue at a ratio of 20:80 (C20S80), and coconut endosperm and soybean curd residue at a ratio of 50:50 (C50S50) without supplementation (numbers after C and S represent their percentage in the formulation; $p < 0.001$). Harvesting was delayed in the BSF fed C80S20 with and without supplementation ($p < 0.001$). The number of total larvae and prepupae was not significantly different between groups ($p > 0.05$). The greatest waste reduction efficiency was observed in the G, C50S50, and C20S80 groups without supplementation ($p < 0.001$). All groups with supplementation had a higher proportion of ash in both larvae and prepupae compared to non-supplemented groups ($p < 0.001$), but lower growth performance. The highest percentage of crude protein in larvae was presented in the Gainesville and C20S80 groups followed by the CK and C50S50 groups ($p < 0.001$). Equal proportions of C and S without supplementation are suggested as a rearing substrate. However, growth performance was

25

lower than for CK; therefore, further studies could investigate cost-efficient techniques to promote this parameter.

Keywords: *Hermetia illucens*; organic waste management; coconut endosperm; soybean curd residue

1. Introduction

The world population has increased sharply in recent decades and could reach 9.7 billion in 2050 [1]. As a consequence, there is a higher demand for food. One-point-three billion tons of food waste is estimated to be generated per year following the sharp increase of consumption [2]. This large amount of organic waste is mainly sent to the landfill [3]. The gas released from this landfill contributes to the greenhouse effect and global warming [4,5]. It is clear from these problems that it is economically sound to study and conduct appropriate management of organic waste.

Several insects have the potential to decompose organic waste and convert it into biomass [6]. One of the most interesting insects to use as a professional decomposer is the black soldier fly (BSF), *Hermetia illucens* (Diptera: Stratiomyidae) [5–7]. BSF is considered a large member of the order Diptera at around 15–20 mm long [8]; it lives in the tropical zone [9], is not a disease vector, and is not harmful to humans or animals [10]. BSF larvae (feeding stage) can efficiently decompose several types of organic waste including poultry manure [11,12], cow manure [12,13], swine manure [12], human feces [13], pig's liver waste [14], fish industrial waste [14], poultry industrial waste [13], restaurant waste [13–15], vegetable waste [13,15], fruit waste [16], vegetable and fruit waste [14,16,17], pineapple and jackfruit peel [18], wheat bran [13,19], maize straw [19], and beer and wine by-products [16]. In addition, BSF larvae can convert this organic waste into high-value products: protein and lipid sources for the animal feed industry [20], biodiesel [21], and antimicrobial peptides [22]. Therefore, BSF rearing could be a solution to managing and upcycling organic waste in an environmentally friendly and economically sustainable way. However, the growth performance, waste reduction efficiency, and chemical composition of BSF are greatly influenced by the rearing substrate [4,5,15]. Based on this knowledge, an appropriate rearing substrate could be used to achieve the highest decomposition efficiency and good-quality end products.

Coconut endosperm (C) and soybean curd residue (S) were used as a rearing substrate in a recent study by Lim et al. [21]; the chemical composition of rearing substrates (crude protein 8.18–20.2% DM (dry matter) and lipid 31.2–31.5% DM) that they reported was quite different from that in our study (crude protein 4.35–11.2% DM and lipid 4.69–5.61% DM), even though the same industrial by-products were used. C and S are easy to obtain in the local market or at industrial scale, as they are considered common by-products. Therefore, C and S were selected for use in this study. In most studies, good chemical composition of BSF was achieved by using chicken diet as a substrate compared to organic waste, but it is not economically sound in reality [15,23] Therefore, supplementation of organic waste with calcium, phosphorus, essential amino acids, and vitamin–mineral premix, to make it similar to chicken diet, could support the performance and quality of BSF. Based on this hypothesis, the objective of this study was to investigate the growth performance, nutritional composition, and waste reduction efficiency of BSF by rearing flies on substrates containing different ratios of C and S, with or without supplementation, compared to the standard diets (Gainesville and starter broiler chicken diet).

2. Materials and Methods

2.1. Insects, Rearing Substrates, and Chemical Analysis

Seven-day-old larvae were randomly collected from a colony of BSF larvae (Orgafeed Co., Ltd., Bangkok, Thailand) which were reared on starter broiler chicken diets (Table 1). All larvae were randomly assigned into eight experimental groups with three replicates per group (2.82 g or approximately 200 larvae per replicate). The larvae in each experimental group were fed different diets: 1. Gainesville diet (G; Scala et al. [5]); 2. Starter broiler chicken diets (CK); 3. Coconut endosperm and soybean curd residue at a ratio of 80:20 (C80S20); 4. Coconut endosperm and soybean curd residue at a ratio of 50:50 (C50S50); 5. Coconut endosperm and soybean curd residue at a ratio of 20:80 (C20S80); 6. Coconut endosperm and soybean curd residue at a ratio of 80:20 with supplementation (C80S20s); 7. Coconut endosperm and soybean curd residue at a ratio of 50:50 with supplementation (C50S50s); 8. Coconut endosperm and soybean curd residue at a ratio of 20:80 with supplementation (C20S80s). The G diet was accepted as the general experimental diet for insects in Diptera and was used as control diet in several research studies [4,5,8,12]. Therefore, the G diet was used as the control diet to compare between groups in this and other studies. The ingredients and chemical composition of diets and supplements were evaluated based on proximate analysis (AOAC 2006) and are shown in Table 1. The supplement was formulated correlating to the macro-minerals, vitamins, micro-minerals, and amino acids in the CK diet. The C and S were obtained as industrial by-products from Dutch Mill Co., Ltd. (Bangkok, Thailand). The C was treated by anaerobic fermentation for 4 weeks prior to usage [21].

Table 1. Ingredients and chemical composition of experimental diets.

Index	Experimental Groups							
	Gainesville diet [1]	Chicken diet [2]	Coconut endosperm/Soybean curd residue					
			80/20	50/50	20/80	80/20	50/50	20/80
			Supplementation [3]					
			No	No	No	Yes	Yes	Yes
Ingredients (%as fed)								
Coconut endosperm	-	-	80	50	20	77.5	47.5	17.5
Soybean curd residue	-	-	20	50	80	17.5	47.5	77.5
Supplements [3]	-	-	-	-	-	5	5	5
Analyzed chemical composition (%Dry matter)								
Dry matter (%Fresh matter)	90.5	88.2	48.4	40.7	32.4	48.8	40.8	31.9
Ash	12.3	6.40	0.70	1.15	1.85	0.68	1.13	1.90
Crude protein	17.1	21.8	4.46	6.90	11.0	4.35	6.91	11.2
Ether extract	1.70	8.00	5.60	5.27	4.73	5.61	5.28	4.69
Crude fiber	10.3	3.40	8.52	7.55	5.94	8.57	7.52	5.82
Nitrogen free extract (NFE)	52.6	60.4	31.1	29.7	27.5	31.1	29.9	27.3
Crude protein/NFE	0.33	0.36	0.14	0.23	0.40	0.14	0.23	0.40

[1] Gainesville diet contains wheat bran, alfalfa meal, and corn meal at 50%, 30%, and 20%, respectively. [2] Chicken diet (Starter) contains corn meal, soybean meal, palm oil, monodicalcium phosphate, limestone, salt, vitamin–mineral premix (Feed specialties Co., Ltd.; Pathumthani, Thailand), DL-methionine, sodium bicarbonate, and choline chloride at 50.8%, 39.3%, 6.06%, 1.38%, 1.38%, 0.24%, 0.36%, 0.25%, 0.20%, and 0.07%, respectively. [3] Five grams of supplement contains monodicalcium phosphate, lime stone, DL-methionine, L-lysine, L-threonine, lard, and vitamin–mineral premix (Feed specialties Co., Ltd.; Pathumthani, Thailand) at 2.2 g, 1 g, 0.3 g, 0.14 g, 0.06 g, 1 g, and 0.3 g, respectively. Vitamin–mineral premix (Feed specialties Co., Ltd.; Pathumthani, Thailand) were supplied per kilogram of diets at 2,500,000 IU of vitamin A; 1,000,000 IU of vitamin D3; 7000 IU of vitamin E; 700 mg of vitamin K; 400 mg of vitamin B1; 800 mg of vitamin B2; 400 mg of vitamin B6; 4 mg of vitamin B12; 30 mg of biotin; 3111 mg of Ca pantothenate acid; 100 mg of folic acid; 15,000 mg of vitamin C; 5600 mg of vitamin B3, 10,500 mg of Zn, 10,920 mg of Fe; 9960 mg of Mn; 3850 mg of Cu; 137 mg of I; 70 mg of Se.

2.2. Rearing, Data Collection, Chemical Analysis, and Calculation

The larvae were placed in a plastic container (15 cm × 24.5 cm × 6.5 cm) on the rearing substrates which were adjusted to obtain an equal humidity of 70% by analyzing the moisture in each substrates and adding water to reach the equal humidity based on calculation before providing into the container. Controlled temperature (28 ± 2 °C) and a dark room were used in this study. Each rearing container was checked twice daily at 09:30 and 16:30. Diet was added into the rearing container to achieve sufficient diet during the experiment. Each replicate was terminated when 40% of the larvae had developed into the prepupal stage [14,19]. The amount of substrate added, amount of substrate remaining, larval weight, prepupal weight, number of larvae, and number of prepupae were measured. Moreover, the pH of rearing substrates was measured at the beginning and end of the experiment by mixing rearing substrates with distilled water at 1:10 w/v [21]. The remaining substrate, larvae and prepupae were frozen and kept at $-20\,^{\circ}$C for further analysis. The larvae and prepupae were dried at 60 $^{\circ}$C for 48 h and ground into a powder by passing through a 1-mm sieve to identify the DM, crude protein (CP), ash, and ether extract (EE), whereas the remaining rearing substrate was evaluated by DM (AOAC 2006). Substrate reduction (%SR), waste reduction index (WRI), and efficiency of conversion of digested food (ECD) were calculated following Meneguz et al. [16] and represented the formulation as below. Larval weight gain was calculated by dividing the increment in total larval weight between 7 and 14 days by seven. This study was carried out following the standard guidelines approved by the Institutional Animal Care and Use Committee of Kasetsart University, Bangkok, Thailand (ACKU63-VET-004).

$$\text{\%SR} = \frac{\text{Distributed substrate(g)} - \text{Residual substrate(g)}}{\text{Distributed substrate(g)}} \times 100 \tag{1}$$

$$\text{WRI} = \frac{\left\{\frac{\text{Distributed substrate(g)} - \text{Residual substrate(g)}}{\text{Distributed substrate(g)}} \times 100\right\}}{\text{Days of trial(day)}} \tag{2}$$

$$\text{ECD} = \frac{\text{Larval and prepupae weight(g)}}{\text{Distributed substrate(g)} - \text{Residual substrate(g)}} \tag{3}$$

2.3. Statistical Analysis

This experiment was performed under a completely randomized design. One-way analysis of variance (ANOVA) was performed to evaluate the differences in all measured, analyzed, and calculated data between experimental groups (fixed factors) by using Duncan's multiple range test as post-hoc analysis. The normal distribution and homogeneity of variance were confirmed by the Shapiro–Wilk test and Levene's test, respectively. Statistically significant difference was accepted at $p < 0.05$. All statistical analyses in the study were investigated by using the R statistics program: RStudio v1.4.1103 with the Rcmdr package (R Development Core Team 2008).

3. Results

The growth performance, chemical composition, waste reduction efficiency, and rearing substrate pH of BSF reared on mixed industrial by-products compared with G and CK are presented in Table 2. The highest larval weight at 14 days and larval growth rate were observed in the CK group followed by the C20S80 and C50S50 groups; the lowest performance was found in the C80S20s group ($p < 0.001$). The latest harvesting date was found for the larvae fed C and S at a ratio of 80:20 with and without supplementation; the harvesting period was around 10–11 days for other groups ($p < 0.001$). The C80S20s group had the lowest final total larval weight ($p < 0.05$). The lowest weight of each larva was presented in the C80S20s group ($p < 0.001$). There was no statistically significant difference in the total number of larvae and prepupae between groups ($p = 0.08$); however, the number of prepupae in the CK and C20S80 groups was higher than in other groups ($p < 0.001$). All groups fed C/S without supplementation (C80S20, C50S50 and C20S80) as well as the C80S20s and G groups had a higher %SR than the CK and C50S50s groups; the lowest was presented in C20S80s ($p < 0.001$). In addition, the C50S50, C20S80, and G groups had a higher WRI than the CK, C80S20, C80S20s, and C50S50 groups; the lowest was presented in C20S80s ($p < 0.001$). The greatest ECD was found in C20S80s followed by C50S50, C80S20, C20S80, C50S50s, CK, G, and C80S20s, respectively. There was a large variation in substrate pH between the beginning of the experiment and the harvesting date: 4.05–6.42 and 4.78–7.17, respectively. The lowest substrate pH at the beginning was presented in C80S20, followed by C50S50, C80S20s, C20S80, C50S50s, C80S20s, G and CK groups, respectively. Substrate pH at the end of the experiment was higher than at the beginning. The most basic substrate condition at the end of the experiment was observed in the G group followed by the CK group compared to the others ($p < 0.001$).

Table 2. Growth performances, waste reduction efficiency, and rearing substrate pH of black soldier fly reared on mixed industrial by-products comparing with Gainesville and chicken diet.

Parameters [1]	Experimental Groups								SEM	*p*-Value
	Gainesville Diet	Chicken Diet	Coconut Endosperm/Soybean Curd Residue							
			80/20	50/50	20/80	80/20	50/50	20/80		
			Supplementation [2]							
			No	No	No	Yes	Yes	Yes		
Growth performances										
Larval weight at 7 days (g) [3]	2.82	2.82	2.82	2.82	2.82	2.82	2.83	2.82	0.002	0.87
Larval weight at 14 days (g) [3]	22.7 [c]	43.4 [e]	17.4 [b]	27.7 [d]	28.2 [d]	12.0 [a]	17.9 [b]	22.4 [c]	1.896	<0.001
Larval weight gain (g/day) [3,4]	2.85 [c]	5.80 [e]	2.08 [b]	3.56 [d]	3.62 [d]	1.31 [a]	2.15 [b]	2.79 [c]	0.271	<0.001
Duration from start to harvesting (days)	10 [a]	10 [a]	15 [c]	10 [a]	10 [a]	16 [c]	11 [b]	11 [b]	0.466	<0.001
At the end of each group [3]										
Final fresh larval weight (g)	25.5 [b,c]	21.0 [a,b,c]	27.5 [c]	28.3 [c]	21.3 [a,b,c]	15.0 [a]	18.4 [a,b]	27.7 [c]	1.228	0.02
Final fresh larval weight (mg/larva)	159 [c,d]	173 [d]	138 [b,c]	152 [c,d]	161 [c,d]	84.2 [a]	122 [b]	148 [b,c,d]	6.110	<0.001
Final fresh prepupal weight (g)	6.93 [a,b]	23.1 [d]	4.95 [a]	10.3 [b]	15.7 [c]	4.60 [a]	7.38 [a,b]	9.54 [b]	1.262	<0.001
Final fresh prepupal weight (mg/prepupa)	100 [a,b]	157 [c]	110 [a,b]	126 [b]	124 [a,b]	97.4 [a]	124 [a,b]	162 [c]	5.305	<0.001
Number of larvae	159	122	200	187	132	173	152	187	7.285	0.05
Number of prepupae	69 [a]	147 [b]	45 [a]	81 [a]	127 [b]	47 [a]	59 [a]	59 [a]	7.649	<0.001
Number of larvae and prepupae	228	269	245	268	259	220	211	246	6.029	0.08
Waste reduction efficiency										
Substrate reduction (%)	49.4 [c]	30.9 [b]	43.9 [c]	43.7 [c]	49.3 [c]	51.4 [c]	28.9 [b]	19.5 [a]	2.448	<0.001
Waste reduction index (g/d) [3]	4.94 [c]	3.09 [b]	2.92 [b]	4.37 [c]	4.93 [c]	3.28 [b]	2.63 [b]	1.77 [a]	0.235	<0.001
Efficiency of conversion of digested food	0.14 [a,b]	0.17 [a,b,c]	0.28 [b,c]	0.31 [c]	0.25 [b,c]	0.07 [a]	0.23 [b,c]	0.56 [d]	0.032	<0.001
pH of feed at										
Beginning of experiment	5.77 [f]	6.42 [g]	4.05 [a]	4.21 [b]	4.82 [c]	4.79 [c]	4.94 [d]	5.36 [e]	0.153	<0.001
Harvesting day	7.17 [d]	6.67 [c]	4.78 [a]	4.97 [a]	5.43 [b]	4.99 [a]	5.10 [a,b]	5.47 [b]	0.174	<0.001

[1] The differences on superscripts in the same row represent the statistical significant difference at $p < 0.05$. [2] Each diet contains five grams of supplement comprising monodicalcium phosphate, lime stone, DL-methionine, L-lysine, L-threonine, lard, and vitamin–mineral premix (Feed specialties Co., Ltd.; Pathumthani, Thailand) at 2.2 g, 1 g, 0.3 g, 0.14 g, 0.06 g, 1 g, and 0.3 g, respectively. [3] Total larvae and/or prepupae were used to calculate these parameters in each study group. [4] Larval weight gain in fresh matter was calculated between 7- to 14 day-old larvae. Vitamin–mineral premix (Feed specialties Co., Ltd.; Pathumthani, Thailand) were supplied per kilogram of diets at 2,500,000 IU of vitamin A; 1,000,000 IU of vitamin D3; 7000 IU of vitamin E; 700 mg of vitamin K; 400 mg of vitamin B1; 800 mg of vitamin B2; 400 mg of vitamin B6; 4 mg of vitamin B12; 30 mg of biotin; 3111 mg of Ca pantothenate acid; 100 mg of folic acid; 15,000 mg of vitamin C; 5600 mg of vitamin B3, 10,500 mg of Zn, 10,920 mg of Fe; 9960 mg of Mn; 3850 mg of Cu; 137 mg of I; 70 mg of Se.

The chemical composition of BSF larvae and prepupae reared on mixed industrial by-products compared with G and CK is presented in Figure 1 and Supplementary Table S1. Based on the percentages of nutrients, ash content was lower in all groups fed C and S without supplementation, in both larvae and prepupae, compared to other groups ($p < 0.001$). The highest ash percentage was present in the groups fed C and S with supplementation; the highest was observed in the C20S80s group ($p < 0.001$), similar to that in CK and G groups. A high CP content in larvae and prepupae was observed in the G and C20S80 groups followed by the C50S50 and CK groups, whereas in other groups, it was lower. The lowest fat proportion in larvae and prepupae was present in the G group ($p < 0.001$). In contrast, the highest fat composition was present in the C80S20 group in both larvae and prepupae ($p < 0.001$).

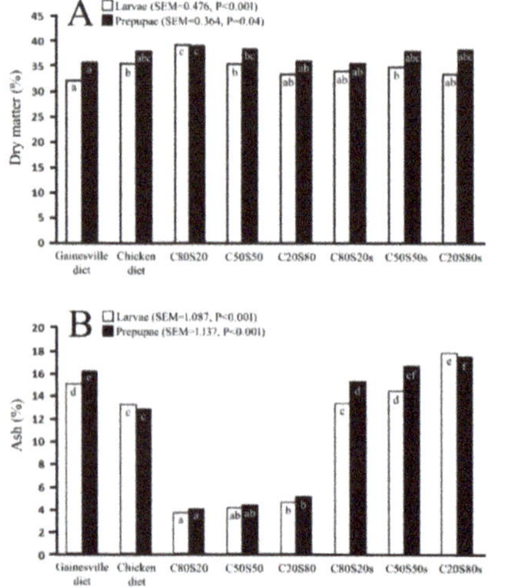

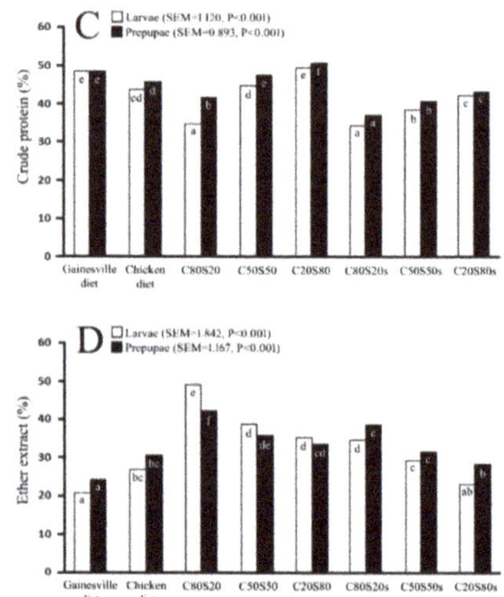

Figure 1. Dry matter ((**A**), %FM), ash ((**B**), %DM), crude protein ((**C**), %DM), and ether extract ((**D**), %DM) of black soldier fly larvae and/or prepupae reared on mixed industrial by-products (C: Coconut endosperm; S: soybean curd residue; number after the abbreviation represents the ratio of mixed industrial by-products) with or without supplementation (s) comparing with Gainesville and chicken diet. The statistical significant difference ($p < 0.05$) is represented by the difference of small capital letters inside bar between the experimental groups of larvae (black color) or prepupae (white color).

4. Discussion

The chemical composition of rearing substrates is considered a major factor influencing the variation in growth performance, waste reduction efficiency, and nutritional composition of BSF larvae and prepupae [14,16,21,24,25]. Lim et al. [21] determined that the total final weight of larvae and growth rate were positively correlated with the amount of CP in the rearing substrate; the highest growth performance was present in larvae fed a mixture of C and S (60:40) containing 12.44% CP. The results of this study agree with those of Lim et al. [21], because the CK diet containing a high CP content (21.8%) provided a higher larval weight at 14 days and greater larval weight gain than in groups fed C and S (4.35–11.2%). Interestingly, Lim et al. [21] found that an excess CP level in the rearing substrate led to negative outcomes for these parameters, which was also represented in this study, because excess protein intake results in energy loss from metabolism involving the excretion of toxic nitrogenous waste [25]. Interestingly, the different chemical composition of the same raw materials between this study and Lim et al.'s [21] was present, but the consequences on growth performances and larval weight gain were similar. Therefore, the cause of this consequence could be influenced by several factors which could be interesting to study. However, CP content could not be the single factor which influences growth parameters. Poor BSF growth performance is found when using rearing substrates with a low caloric density, i.e., low fat and/or carbohydrate, because these nutrients serve as energy sources [23]. Generally, an appropriate ratio between CP and metabolizable energy must be formulated to obtain the highest performance in livestock animals [4]. Therefore, appropriate chemical composition of the rearing substrate could be another consideration point for BSF as it is in livestock animals. A 1:1 ratio of CP to carbohydrate for BSF cultivation has been reported to achieve the fastest development [4]. In our study, the CP:carbohydrate ratio of 0.36 in the CK diet provided the highest performance. On the one hand, C50S50 and C20S80 with CP:carbohydrate ratios of 0.23 and 0.40, respectively, presented a higher larval growth rate compared to other groups fed industrial by-products. Poor performance was found in this study when the CP:carbohydrate ratio was 0.14. However, it still cannot be concluded that only this ratio is the primary factor influencing growth rate. In our aspect, nutrient quality of amino acids and digestibility could be further studied in depth as another consideration point.

There was no significant difference in the total number of larvae and prepupae between groups in this study. We can assume that the mortality rate was not affected by experimental diets. Most studies report that the mortality rate does not change between rearing diets [21], because BSF can survive on poor nutrient diets and large environmental condition [26]. However, the duration from larvae to prepupae is prolonged when larvae are reared on diets lacking certain nutrients, mainly protein, which prolongs the cultivation period [24,25]. The feeding period is prolonged until the nutrients inside the larvae meet the requirements for development and metamorphosis at which stage they can no longer consume feed [24]. The latest harvesting date (22–23 days old) was found in the larvae fed C and S at a ratio of 80:20, which is a low protein proportion compared to that fed to other groups (which were harvested at 17 days old). In another study, the shortest rearing period (19 days old) was presented by larvae fed a mixture of C and S at a ratio of 60:40, containing the highest CP compared to other study groups [21]. In this study, the larvae being fed the CK diet for 7 days before consuming the organic waste could be the cause of an earlier harvesting date compared to the study of Lim et al. [21], in which larvae were reared on organic waste throughout the experiment [13,20]. Therefore, an appropriate starter diet could be used before rearing on organic waste to reduce rearing duration.

The pH condition of substrates did not influence the final weight [27]. However, Ma et al. [28] demonstrated that the initial substrate pH influences the final weight, the best growth performance being observed at pH 6. In the same way, Lim et al. [21] found the highest total weight and growth rate when rearing BSF at pH 5.82. In addition, it has been suggested that the rearing substrate pH should be higher than 6 to achieve a good productive performance [9]. In our study, the substrate with a pH of 6.42 (CK diet) resulted in the significantly heaviest larvae weight at 14 days, whereas a lower larvae weight was found in other groups fed industrial by-products with an initial substrate pH of 4.05–5.36. Therefore, the difference in initial substrates pH could be another cause of the diverse outcomes in this study.

The rearing substrate has a direct impact on the nutritional composition of BSF larvae and prepupae [15,16,25]. The chemical composition of the mixed organic diet and that of the BSF larvae and prepupae in this study demonstrate that the CP content in the rearing diet influences BSF larval and prepupal protein content. A high CP content in the diet produced a high protein content in larvae, a result similar to that of other studies [16]. Nitrogen-free extract (NFE) in the diet is positively correlated to the fat content in BSF larvae [25]. In this study, the increment of NFE in the mixed organic diet promoted a higher fat content in BSF larvae than in another study [15]. In addition, larvae reared on a diet high in soluble carbohydrates (618 g/kg) had the highest fat content (386 g/kg DM), whereas those reared on a diet low in soluble carbohydrates (7 g/kg) had the lowest fat content (218 g/kg DM), because insect larvae can convert excess carbohydrates to fat and store it in their body mass [16].

A great increase of ash composition (around three times) was observed in this study in both larvae and prepupae fed supplemented industrial by-products compared to the non-supplemented group. Increasing the calcium level in the substrate correlates with a higher calcium composition in house cricket (*Acheta domestica*) and yellow mealworm [29]. The supplement mainly contained calcium and phosphorus, which could be the cause of this consequence. Therefore, increasing the mineral content in the rearing substrate promoted an increase of these minerals in BSF larvae and prepupae as it did in house cricket and yellow mealworm [29]. From current knowledge from our and other studies, CK diet promotes good growth performance for BSF [15]. Therefore, the supplement was formulated to supply calcium, phosphorus, amino acids, and vitamin–mineral premix to industrial by-product diets, reflecting the CK diet which can provide benefits. However, the nutrient profiles of CK and industrial by-product diets are quite different. Our results showed that the ash content in BSF larvae and prepupae fed with supplements was similar to that in those fed CK, but there was a deterioration in growth performance compared to CK. In addition, major nutrients such as CP and carbohydrate could be more important factors influencing the growth performance than the supplement. Therefore, increasing the mineral content in rearing substrates which do not contain as appropriate a nutritive value as CK is not indicated for BSF, because a deterioration in growth performance can occur as in this study.

5. Conclusions

In this study, our results for growth performance, waste reduction efficiency, and the nutritional composition of BSF larvae reared on C50S50 were very similar to those for BSF larvae reared on standard diets (G and CK). However, the growth performance remained lower than in BSF fed CK. Supplementation is not necessary to improve growth performance, waste reduction efficiency, and nutritional composition when feeding larvae industrial by-product diets. Further studies could investigate the clear nutrient requirements of BSF, correlating them with their performance and quality. A technique for improving the growth performance of BSF, to reach a level similar to or better than that of larvae fed CK, by using industrial by-products as rearing substrates, could also be studied.

Supplementary Materials: The following are available online at https://www.mdpi.com/article/10.3390/insects12080682/s1, Table S1: Chemical composition of black soldier fly larvae and prepupae reared on mixed industrial by-products comparing with Gainesville and chicken diet.

Author Contributions: Conceptualization, A.K.; methodology, N.P. and A.K.; formal analysis, N.P. and A.K.; investigation, N.P., P.C. and A.K.; data curation, N.P. and A.K.; writing—original draft preparation, N.P. and A.K.; writing—review and editing, N.P., P.C. and A.K.; project administration, A.K.; funding acquisition, A.K. All authors have read and agreed to the published version of the manuscript.

Funding: This research was funded by Thailand Science Research and Innovation and Office of National Higher Education Science and Research and Innovation Policy Council (Thailand). This work was partially supported by the Faculty of Veterinary Medicine, Kasetsart University, Bangkok, Thailand.

Institutional Review Board Statement: Not applicable.

Acknowledgments: The authors would like to humbly thank Orgafeed Co. Ltd. (Bangkok, Thailand) and Dutch Mill Co., Ltd. (Bangkok, Thailand) for providing insects and rearing substrates. The authors would like to thank Sathita Areerat and Penpicha Kongsup for technical assistance on animal husbandry and sample collection.

Conflicts of Interest: The authors declare no conflict of interest.

References

1. Growing at a Slower Pace, World Population Is Expected to Reach 9.7 Billion in 2050 and Could Peak at Nearly 11 Billion around 2100. Available online: https://www.un.org/development/desa/en/news/population/world-population-prospects-2019.html (accessed on 19 October 2020).
2. Food Wastage Footprint. *Full-Cost Accounting*; Final Report; Food and Agriculture Organization of the United Nations: Rome, Italy, 2014; p. 10.
3. Grossule, V.; Vanin, S.; Lavagnolo, M.C. Potential treatment of leachate by *Hermetia illucens* (Diptera, Stratyomyidae) larvae: Performance under different feeding conditions. *Waste Manag.* **2020**, *38*, 537–545. [CrossRef]
4. Cammack, J.A.; Tomberlin, J.K. The Impact of Diet Protein and Carbohydrate on Select Life-History Traits of The Black Soldier Fly *Hermetia illucens* (L.) (Diptera: Stratiomyidae). *Insects* **2017**, *8*, 56. [CrossRef] [PubMed]
5. Scala, A.; Cammack, J.A.; Salvia, R.; Scieuzo, C.; Franco, A.; Bufo, S.A.; Tomberlin, J.K.; Falabella, P. Rearing substrate impacts growth and macronutrient composition of *Hermetia illucens* (L.) (Diptera: Stratiomyidae) larvae produced at an industrial scale. *Sci. Rep.* **2020**, *10*, 19448. [CrossRef] [PubMed]
6. Cickova, H.; Newton, G.L.; Lacy, R.C.; Kozanek, M. The use of fly larvae for organic waste treatment. *Waste Manag.* **2015**, *35*, 68–80. [CrossRef]
7. Surendra, K.C.; Tomberlin, J.K.; van Huis, A.; Cammack, J.A.; Heckmann, L.L.; Khanal, S.K. Rethinking organic wastes bioconversion: Evaluating the potential of the black soldier fly (*Hermetia illucens* (L.)) (Diptera: Stratiomyidae) (BSF). *Waste Manag.* **2020**, *117*, 58–80. [CrossRef] [PubMed]

8. Sheppard, D.C.; Tomberlin, J.K.; Joyce, J.A.; Kiser, B.C.; Sumner, S.M. Rearing methods for the black soldier fly (Diptera: Stratiomyidae). *Med. Entomol.* **2002**, *39*, 695–698. [CrossRef]

9. Singh, A.; Kumari, K. An inclusive approach for organic waste treatment and valorisation using Black Soldier Fly larvae: A review. *J. Environ. Manag.* **2019**, *251*, 109569. [CrossRef] [PubMed]

10. Wang, Y.S.; Shelomi, M. Review of Black Soldier Fly (*Hermetia illucens*) as Animal Feed and Human Food. *Foods* **2017**, *6*, 91. [CrossRef]

11. Gutiérrez, G.; Ruiz, R.; Vélez, H. Compositional, microbiological and protein digestibility analysis of the larva meal of *Hermetia illuscens* L. (Diptera:Stratiomyiidae) at angelópolis–antioquia, colombia. *Rev. Fac. Nac. Agron. Medellín* **2004**, *57*, 2491–2500.

12. Zhou, F.; Tomberlin, J.; Zheng, L.; Ziniu, Y.; Zhang, J. Developmental and Waste Reduction Plasticity of Three Black Soldier Fly Strains (Diptera: Stratiomyidae) Raised on Different Livestock Manures. *Med. Entomol.* **2013**, *50*, 1224–1230. [CrossRef]

13. Gold, M.; Cassar, C.M.; Zurbrugg, C.; Kreuzer, M.; Boulos, S.; Diener, S.; Mathys, A. Biowaste treatment with black soldier fly larvae: Increasing performance through the formulation of biowastes based on protein and carbohydrates. *Waste Manag.* **2020**, *102*, 319–329. [CrossRef]

14. Nguyen, T.T.; Tomberlin, J.K.; Vanlaerhoven, S. Ability of Black Soldier Fly (Diptera: Stratiomyidae) Larvae to Recycle Food Waste. *Environ. Entomol.* **2015**, *44*, 406–410. [CrossRef]

15. Spranghers, T.; Ottoboni, M.; Klootwijk, C.; Ovyn, A.; Deboosere, S.; Meulenaer, B.; Michiels, J.; Eeckhout, M.; De Clercq, P.; De Smet, S. Nutritional composition of black soldier fly (*Hermetia illucens*) prepupae reared on different organic waste substrates. *J. Sci. Food Agric.* **2017**, *97*, 2594–2600. [CrossRef] [PubMed]

16. Meneguz, M.; Schiavone, A.; Gai, F.; Dama, A.; Lussiana, C.; Renna, M.; Gasco, L. Effect of rearing substrate on growth performance, waste reduction efficiency and chemical composition of black soldier fly (*Hermetia illucens*) larvae. *J. Sci. Food Agric.* **2018**, *98*, 5776–5784. [CrossRef]

17. Giannetto, A.; Oliva, S.; Ceccon Lanes, C.F.; de Araujo Pedron, F.; Savastano, D.; Baviera, C.; Parrino, V.; Lo Paro, G.; Spano, N.C.; Cappello, T.; et al. *Hermetia illucens* (Diptera: Stratiomydae) larvae and prepupae: Biomass production, fatty acid profile and expression of key genes involved in lipid metabolism. *J. Biotechnol.* **2020**, *307*, 44–54. [CrossRef] [PubMed]

18. Burana, K.; Jamjanya, T. Population Size Monitoring, Rearing Methods and Nutritional Value of Black Soldier Fly (*Hermetia illucens* L.). *KKU Res. J.* **2011**, *11*, 19–26. [CrossRef]

19. Gao, Z.; Wang, W.; Lu, X.; Zhu, F.; Liu, W.; Wang, X.; Lei, C. Bioconversion performance and life table of black soldier fly (*Hermetia illucens*) on fermented maize straw. *J. Clean. Prod.* **2019**, *230*, 974–980. [CrossRef]

20. Ewald, N.; Vidakovic, A.; Langeland, M.; Kiessling, A.; Sampels, S.; Lalander, C. Fatty acid composition of black soldier fly larvae (*Hermetia illucens*)–Possibilities and limitations for modification through diet. *Waste Manag.* **2020**, *102*, 40–47. [CrossRef]

21. Lim, J.W.; Mohd-Noor, S.N.; Wong, C.Y.; Lam, M.K.; Goh, P.S.; Beniers, J.J.A.; Oh, W.D.; Jumbri, K.; Ghani, N.A. Palatability of black soldier fly larvae in valorizing mixed waste coconut endosperm and soybean curd residue into larval lipid and protein sources. *Waste Manag.* **2019**, *231*, 129–136. [CrossRef]

22. Vogel, H.; Muller, A.; Heckel, D.G.; Gutzeit, H.; Vilcinskas, A. Nutritional immunology: Diversification and diet-dependent expression of antimicrobial peptides in the black soldier fly *Hermetia illucens*. *Dev. Comp. Immunol.* **2018**, *78*, 141–148. [CrossRef]

23. Nguyen, T.T.; Tomberlin, J.K.; Vanlaerhoven, S. Influence of resources on *Hermetia illucens* (Diptera: Stratiomyidae) larval development. *Med. Entomol.* **2013**, *50*, 898–906. [CrossRef]

24. Banks, I.J.; Gibson, W.T.; Cameron, M.M. Growth rates of black soldier fly larvae fed on fresh human faeces and their implication for improving sanitation. *Med. Int. Health* **2014**, *19*, 14–22. [CrossRef]

25. Gold, M.; Tomberlin, J.K.; Diener, S.; Zurbrugg, C.; Mathys, A. Decomposition of biowaste macronutrients, microbes, and chemicals in black soldier fly larval treatment: A review. *Waste Manag.* **2018**, *82*, 302–318. [CrossRef]

26. Lalander, C.; Diener, S.; Zurbrügg, C.; Vinnerås, B. Effects of feedstock on larval development and process efficiency in waste treatment with black soldier fly (*Hermetia illucens*). *J. Clean. Prod.* **2019**, *208*, 211–219. [CrossRef]

27. Meneguz, M.; Gasco, L.; Tomberlin, J.K. Impact of pH and feeding system on black soldier fly (*Hermetia illucens*, L.; Diptera: Stratiomyidae) larval development. *PLoS ONE* **2018**, *13*, e0202591. [CrossRef] [PubMed]

28. Ma, J.; Lei, Y.; Rehman, K.U.; Yu, Z.; Zhang, J.; Li, W.; Li, Q.; Tomberlin, J.K.; Zheng, L. Dynamic Effects of Initial pH of Substrate on Biological Growth and Metamorphosis of Black Soldier Fly (Diptera: Stratiomyidae). *Environ. Entomol.* **2018**, *47*, 159–165. [CrossRef] [PubMed]

29. Anderson, S.J. Increasing Calcium Levels in Cultured Insects. *Zoo Biol.* **2000**, *19*, 1–9. [CrossRef]

Article

Effects of Different Nitrogen Sources and Ratios to Carbon on Larval Development and Bioconversion Efficiency in Food Waste Treatment by Black Soldier Fly Larvae (*Hermetia illucens*)

Yan Lu [1,2,†], Shouyu Zhang [1,2,†], Shibo Sun [3], Minghuo Wu [1], Yongming Bao [1], Huiyan Tong [1], Miaomiao Ren [1], Ning Jin [1], Jianqiang Xu [2,3], Hao Zhou [1,4] and Weiping Xu [1,2,4,*]

[1] Department of Environmental Ecological Engineering, School of Marine Science and Technology, Dalian University of Technology, Panjin 124221, China; lyllyying@mail.dlut.edu.cn (Y.L.); Zshouyu@mail.dlut.edu.cn (S.Z.); wumh@dlut.edu.cn (M.W.); biosci@dlut.edu.cn (Y.B.); tonghuiyan@dlut.edu.cn (H.T.); renmiao18@mail.dlut.edu.cn (M.R.); 18342782101@163.com (N.J.); zhouhao@dlut.edu.cn (H.Z.)

[2] Panjin Institute of Industrial Technology, Dalian University of Technology, Panjin 124221, China; Jianqiang.Xu@dlut.edu.cn

[3] School of Life Science and Pharmaceutical Sciences, Dalian University of Technology, Panjin 124221, China; sunshibo@mail.dlut.edu.cn

[4] Key Laboratory of Industrial Ecology and Environmental Engineering, Dalian University of Technology, Dalian 116024, China

* Correspondence: Weiping.Xu@dlut.edu.cn

† These authors contributed equally to this study.

Citation: Lu, Y.; Zhang, S.; Sun, S.; Wu, M.; Bao, Y.; Tong, H.; Ren, M.; Jin, N.; Xu, J.; Zhou, H.; et al. Effects of Different Nitrogen Sources and Ratios to Carbon on Larval Development and Bioconversion Efficiency in Food Waste Treatment by Black Soldier Fly Larvae (*Hermetia illucens*). *Insects* **2021**, *12*, 507. https://doi.org/10.3390/insects12060507

Academic Editors: Man P. Huynh, Kent S. Shelby and Thomas A. Coudron

Received: 18 April 2021
Accepted: 11 May 2021
Published: 31 May 2021

Publisher's Note: MDPI stays neutral with regard to jurisdictional claims in published maps and institutional affiliations.

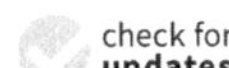

Simple Summary: Black soldier fly larvae (BSFL) have received global research interest and industrial application due to their high performance on the organic waste treatment. However, the substrate C/N property, which may affect larvae development and the waste bioconversion process greatly, is significantly less studied. The current study focused on the food waste treatment by BSFL, compared the nitrogen supplying effects of 9 nitrogen species (i.e., NH_4Cl, $NaNO_3$, urea, uric acid, Gly, L-Glu, L-Glu:L-Asp (1:1, w/w), soybean flour, and fish meal), and further examined the C/N effects on the larval development and bioconversion process. We found that NH_4Cl and $NaNO_3$ led to poor larval growth and survival, while 7 organic nitrogen species exerted no harm to the larvae. Urea was further chosen to adjust the C/Ns. Results showed that lowering the C/N from the initial 21:1 to 18:1–14:1 improved the waste reduction and larvae production performance, and C/N of 18:1–16:1 was further beneficial for the larval protein and lipid bioconversion, whereas C/N of 12:1–10:1 resulted in a significant performance decline. Therefore, the C/N range of 18:1–16:1 is likely the optimal condition for food waste treatment by BSFL and adjusting food waste C/N with urea could be a practical method for the performance improvement.

Abstract: Biowaste treatment by black soldier fly larvae (BSFL, *Hermetia illucens*) has received global research interest and growing industrial application. Larvae farming conditions, such as temperature, pH, and moisture, have been critically examined. However, the substrate carbon to nitrogen ratio (C/N), one of the key parameters that may affect larval survival and bioconversion efficiency, is significantly less studied. The current study aimed to compare the nitrogen supplying effects of 9 nitrogen species (i.e., NH_4Cl, $NaNO_3$, urea, uric acid, Gly, L-Glu, L-Glu:L-Asp (1:1, w/w), soybean flour, and fish meal) during food waste larval treatment, and further examine the C/N effects on the larval development and bioconversion process, using the C/N adjustment with urea from the initial 21:1 to 18:1, 16:1, 14:1, 12:1, and 10:1, respectively. The food wastes were supplied with the same amount of nitrogen element (1 g N/100 g dry wt) in the nitrogen source trial and different amount of urea in the C/N adjustment trial following larvae treatment. The results showed that NH_4Cl and $NaNO_3$ caused significant harmful impacts on the larval survival and bioconversion process, while the 7 organic nitrogen species resulted in no significant negative effect. Further adjustment of C/N with urea showed that the C/N range between 18:1 and 14:1 was optimal for a high waste reduction performance (73.5–84.8%, $p < 0.001$) and a high larvae yield (25.3–26.6%, $p = 0.015$), while the C/N range of 18:1 to 16:1 was further optimal for an efficient larval protein yield (10.1–11.1%, $p = 0.003$) and lipid yield (7.6–8.1%, $p = 0.002$). The adjustment of C/N influenced the activity of antioxidant enzymes, such as superoxide dismutase (SOD, $p = 0.015$), whereas exerted no obvious impact on the larval amino acid composition. Altogether, organic nitrogen is more suitable than NH_4Cl and $NaNO_3$ as the nitrogen amendment during larval food waste treatment, addition of small amounts of urea, targeting C/N of 18:1–14:1, would improve the waste reduction performance, and application of C/N at 18:1–16:1 would facilitate the larval protein and lipid bioconversion process.

Keywords: nitrogen source; carbon to nitrogen ratio; food waste; urea; black soldier fly larvae; *Hermetia illucens*

---

1. Introduction

Larvae of black soldier fly (BSFL), Diptera:Stratiomyidae, *Hermetia illucens* (Linnaeus), are capable of converting various biowaste into protein-rich insect biomass and nitrogen-rich organic fertilizer [1–3] and have received worldwide research interest and fast-growing industrial application [4,5]. The application of BSFL for waste processing has expanded from tropical and temperate countries [4,5] to Russia [6], Canada [7], and Near East Turkey [8] in recent years. Amongst all the parameters that may impact the larval development and biowaste conversion process, ones such as temperature, moisture content, and pH are critically analyzed [9–11]. In contrast, the substrate C/N property, which may play a crucial role in the larval development and bioconversion process, is largely less studied. Several studies have examined the C/N effects indirectly, for instance, Bessigamukama et al. [12], Ewald et al. [13], and Lopes et al. [14] have added different amounts of biochar [12], fish [13], or mussel [14] to the grain or bread substrate for BSFL treatment, and Pang et al. [15] has studied C/N effects on the greenhouse gas emission; however, the C/N effects on larval development are still not clear and the optimal C/N range for a high bioconversion performance has not been achieved, since the substrate total C/N ratios were not determined or the bioconversion efficiencies of BSFL treatment were not examined in those studies.

The BSFL are able to degrade a wide range of organic waste, such as animal manure, food waste, abattoir waste, and aquaculture waste [16–19]. These wastes have different C/N properties, and BSFL are found to be largely more adapted to the C/N range of <20:1 and less suitable to the C/N range of >20:1. For instance, poultry, swine, and human feces, typically having C/N of 9:1 to 15:1, were found to be well-degraded by BSFL, while cow manure, having C/N of 20:1 to 30:1, was less decomposed by BSFL [20,21]. Poultry feed (C/N 18:1), food waste (C/N 14:1), and abattoir waste (C/N 6:1) showed acceptable and comparable decomposition, while fruit and vegetable waste (C/N 24:1) was found to be less processed by the BSFL [18]. In contrast to the animal manure waste, the food waste could be widely different on the C/N properties due to the waste composition. The Chinese diet habit results in a great proportion of carbohydrate (rice, noodles, and steamed buns) left in the food waste, similar to those food wastes containing high amounts of starches [22], which lead the C/N to be occasionally higher than 20:1. This kind of food waste needs to be recycled by the BSFL, which results in an interesting substrate with high C/N properties and arouses practical questions about what nitrogen source should be used for the nitrogen supplement and what C/N range is suitable for a high performance of BSFL treatment.

Both inorganic and organic N species could be examined for the nitrogen supplemental effects on the food waste treatment. Previously, biowaste leachate has been reported to be treated by BSFL [23,24]. Since NH_4-N and NO_3-N are typical nitrogen species in the leachate, the NH_4Cl and $NaNO_3$ could be tested for the nitrogen supplemental performance. Poultry and swine feces are reported to be effectively degraded by BSFL [20,21]. Since urea and uric acid are typical compounds in the feces, these two chemicals could be tested for the nitrogen supplemental efficiency. BSFL protein is found to be rich with L-Glu and

L-Asp amino acids [25]. The L-Glu, L-Asp, as well the simplest amino acid of Gly could be tested for the N supplemental efficiency. Soybean flour and fish meal could be used as positive controls for the organic nitrogen species. As the BSFL are adapted to the poultry manure of C/N 9:1 and less adapted to the cow manure of 20:1 [21,26], the food waste C/N ratio could be adjusted from 20:1 to 10:1 in order to identify the optimal C/N range for a high larval conversion performance.

Therefore, the present study aimed to compare the nitrogen supplying effects of 9 nitrogen species (i.e., NH_4Cl, $NaNO_3$, urea, uric acid, Gly, L-Glu, L-Glu:L-Asp (1:1, w/w), soybean flour, and fish meal) and further examine the C/N effects on the larval development and bioconversion process for the food waste treatment. Larval enzyme activity and amino acid composition were partially analyzed in order to study the potential physiological effects of nitrogen supplement.

2. Materials and Methods

2.1. Larvae, Food Waste, and Nitrogen Source Preparation

Black soldier fly eggs were purchased from a BSF farm (Baiaotai farm, Anyou Biotechnology Group Co., Ltd., Guangxi, China). Upon arrival, eggs were hatched for six days in a substrate containing 60% soybean meal, 30% corn powder, and 10% wheat bran in a 65% moisture content environment at 25 °C. The 6-day old larvae (average weight 0.0027 g) were removed from the hatching substrate through sieving (1 mm mesh) and 45 batches of approximately 800 larvae were weighed.

Food wastes (FW) were the cooked food leftovers (rice, noodles, vegetables, meats, eggs, etc.) that were collected from the university canteen (Dalian University of Technology, Panjin Campus, Panjin, China). After collection, food wastes were homogenized with a kitchen blender, tested for moisture content in duplicate (oven drying at 105 °C until constant weight [27]), and stored at 4 °C or −20 °C prior to further usage. In Trial 1, 9 nitrogen sources were used, including NH_4Cl, $NaNO_3$, urea, uric acid, Gly, L-Glu, L-Asp, soybean flour, and fish meal, while only urea was used in Trial 2. Within the 9 nitrogen sources, 7 pure compounds with purity grade >99% (NH_4Cl, $NaNO_3$, urea, uric acid, Gly, L-Glu, and L-Asp) were purchased from Aladdin (Shanghai Aladdin Biochemical Technology Co., Ltd., Shanghai, China), and 2 nitrogen sources (soybean flour and fish meal) were purchased from local stores. Carbon and nitrogen contents of the 7 pure compounds were calculated based on their molecular weight, while the C and N properties of the soybean flour, fish meal, as well as the food waste were determined in duplication using the Vario EL cube elemental analyzer (Elementar Analysensysteme GmbH, Hanau, Germany) with the freeze-dried subsamples (Table 1).

2.2. Experimental Design

In Trial 1, different nitrogen sources were supplied in 1 g N element per 100 g (dry wt) food waste, i.e., changing N% of food waste from 2.26% to 3.26%. Since the 9 nitrogen sources contained different N and C contents, the actual weights of each nitrogen source used are reported in the Table 1, as well as the C/N values after the nitrogen supplement. In Trial 2, urea was added to the food waste, aiming to adjust food waste C/N from 21:1 (blank control) to 18:1, 16:1, 14:1, 12:1, and 10:1, respectively, while the actual weights of urea used were listed in the Table 1.

All the experiments in Trials 1 and 2 were performed in triplicate, with food waste without nitrogen amendment served as the blank control. Larvae were reared in 4.6 L plastic boxes (240 × 120 × 160 mm) individually. Ten 6 mm diameter holes were made on the box lid in order to enhance passive aeration. In each box, 300 g (wet weight, 70% moisture content) of food waste was added, 150 g food waste was added on Day 0, and another 150 g food waste was added on Day 6. The nitrogen sources were added associated with the food waste according to the amounts in Table 1 on Days 0 and 6, respectively. The 800 weighed larvae (6-days-old) were added into each box on Day 0 following the addition of food waste and nitrogen sources. The boxes were kept at 26–32 °C, and the substrates were mixed manually twice per day. After 12 days, larvae in each box were separated from the frass manually, rinsed with tap water, and dried on paper towels. The total number of larvae was counted. The total wet weight of larvae and frass were recorded, and the moisture content of subsamples of larvae and frass were determined by oven-drying at 105 °C until constant weight [27]. The total dry content of larvae and frass were calculated. Subsamples of larvae were used for the protein and lipid content determination as well as enzyme activity analysis. The rest of the samples were stored at −20 °C for further analysis.

Table 1. Experimental design of nitrogen source supplied in Trials 1 and 2.

Groups	N Source Supplied [1]	N Element Amount	N Source Amount	Final Carbon and Nitrogen Properties after Nitrogen Source Supplement		
		g/100 g FW	g/100 g FW	C (%)	N (%)	C/N
Trial 1						
Blank	None	0	0	47.0	2.26	21:1
NH_4CL	NH_4CL	1	3.82	47.0	3.26	14:1
$NaNO_3$	$NaNO_3$	1	6.06	47.0	3.26	14:1
Urea	Urea	1	2.14	47.4	3.26	15:1
Uric acid	Uric acid	1	3.00	48.1	3.26	15:1
Gly	Gly	1	5.35	48.7	3.26	15:1
Glu	L-Glu	1	10.53	51.3	3.26	16:1
Glu/Asp	L-Glu/L-Asp (1:1)	1	10.00	50.9	3.26	16:1
Soybean flour	Soybean flour	1	16.72	55.1	3.26	17:1
Fish meal	Fish meal	1	9.52	50.8	3.26	16:1
Trial 2						
Blank (21:1)	None	0	0	47.0	2.26	21:1
C/N(18:1)	Urea	0.36	0.76	47.2	2.62	18:1
C/N(16:1)	Urea	0.68	1.46	47.3	2.94	16:1
C/N(14:1)	Urea	1.10	2.36	47.5	3.36	14:1
C/N(12:1)	Urea	1.66	3.55	47.7	3.92	12:1
C/N(10:1)	Urea	2.44	5.23	48.0	4.70	10:1

[1] The C and N contents were determined for food waste (FW, C 47.0%, N 2.26%, $n = 2$), soybean flour (C 48.5%, N 5.98%, $n = 2$), and fish meal (C 39.9%, N 10.5%, $n = 2$) materials. All parameters were obtained on a dry matter basis.

2.3. Analysis of Larval Development and Nutrient Composition

Larval length and weight were determined with a two-day interval over 12 days of treatment. Body length was determined in triplicates for larvae within the same box, and the body weight was measured by a combined 10-larva weight that was averaged to obtain the single weight. Larval protein and lipid contents were determined for samples collected on Day 12. To determine protein content, larvae were freeze-dried and milled, larval C and N contents were measured with the Vario EL cube elemental analyzer, and larval protein contents were calculated as the nitrogen content × 4.67 following Janssen et al. [28]. In the lipid analysis, freeze-dried and milled larvae were extracted by petroleum ether (Aladdin) twice (1:10, *w/v*, 48 h, 25 °C) in order to achieve crude lipid according to Zheng et al. [29]. After evaporating the petroleum ether, the crude lipid weights were recorded, and the lipid contents were measured as the ratio of crude lipid weight to the larval dry matter. Furthermore, the larval total protein and lipid yield in g/100 g dry waste were evaluated based on the equations below and as mentioned previously [13,30,31].

$$Protein\ yield\ \% \ = \frac{L \times protein\%}{W} \times 100 \tag{1}$$

$$Lipid\ yield\ \% = \frac{L \times lipid\%}{W} \times 100 \tag{2}$$

where L represents the total dry matter of larvae on Day 12, protein% represents larval protein content (Day 12), lipid% represents larval lipid content (Day 12), and W represents the total dry matter of food waste, including the nitrogen supplement. All parameters were obtained in grams on a dry matter basis.

2.4. Assessment of the Process Efficiency

To assess the larval treatment efficiency, the larval survival ratio (SR), waste reduction ratio (WR), larvae yield (LY), nitrogen conversion ratio (NCR), efficiency of conversion of digested feed (ECD), as well as the mass distribution pattern were evaluated by applying the aforementioned equations [14,18,30,31]:

$$Survival\ ratio\ (SR)\ \% \ = \frac{Larvae_{end}}{Larvae_{beg}} \times 100 \tag{3}$$

$$Waste\ reduction\ ratio\ (WR)\ \% = \frac{W - R}{W} \times 100 \tag{4}$$

$$Larvae\ yield\ (LY)\ \% = \frac{L}{W} \times 100 \tag{5}$$

$$Nitrogen\ conversion\ ratio\ (NCR)\ \% = \frac{L \times N\%_{larvae}}{W \times N\%_{waste}} \times 100 \tag{6}$$

$$Efficiency\ of\ conversion\ (ECD)\ \% = \frac{L}{W - R} \times 100 \tag{7}$$

$$Mass\ balance:\ W = R + L + M \tag{8}$$

where $Larvae_{beg}$ and $Larvae_{end}$ represent the larval numbers at the beginning and end of the treatment respectively, W, R, and L represent the total dry matter of food waste including the nitrogen supplement (W), the frass residue (R), and the larvae (L) respectively, M represents the dry matter loss as a result of larval and microbial metabolism (M), $N\%_{larvae}$ represents the nitrogen content of larvae on Day 12, and $N\%_{waste}$ represents the nitrogen content of food waste, including the nitrogen supplement. All parameters were obtained in grams on a dry matter basis.

2.5. Enzyme Activity and Amino Acid Composition Analysis

Following C/N adjustment with urea in Trial 2, the larval enzyme activity and amino acid composition were further analyzed in triplicate. The activities of antioxidant enzymes, such as peroxidase (POD), superoxide dismutase (SOD), catalase (CAT), and glutathione peroxidase (GSH-px), were observed and recorded. Fresh larvae collected on Day 12 were homogenized by a $1\times$ phosphate-buffered solution (PBS, pH 7.4) in a 1:10 (w/v) ratio and then centrifuged at $10,000\times g$ for 3 min. The supernatants were collected and analyzed for the POD, SOD, CAT, and GSH-px activity following the practice of Chen et al. [32]. Briefly, the POD activity was determined at 420 nm using substrate containing H_2O_2, and 1 U of POD was defined as the amount of enzyme that catalyzed 1 μg substrate per liter per minute. The SOD activity was determined at 550 nm through the xanthine and xanthine oxidase system, and 1 U of SOD was defined as the amount of enzyme that created 50% inhibition of xanthine oxidase. The CAT activity was analyzed by measuring the absorbance decrease at 240 nm due to H_2O_2 decomposition, and 1 U of CAT was defined as the amount of enzyme that decomposed 1 μmol H_2O_2 per liter per second. The GSH-px activity was measured at 423 nm in a system containing 5,5-dithio-bis-(2-nitrobenzoic acid) (DTNB) and reduced GSH, and 1 U of GSH-px was defined as the amount of enzyme that oxidized 1 μmol of reduced GSH per liter per minute. All enzyme activities were calculated and recorded in the unit of U/g larval wet weight.

As for the amino acid composition analysis, freeze-dried and milled larval samples were oxidized by performic acid at 4 °C for 16 h and then hydrolyzed by 6 M HCl with 0.1% phenol at 110 °C for 22 h. The hydrolyzed aliquots were then diluted by 0.02 M HCl and analyzed using a L-8900 High-speed Amino Acid Analyzer (Hitachi High-Tech, Japan) following the instruction manual. For Trp (trytophan), the freeze-dried and milled larval samples were hydrolyzed by 5 M NaOH at 110 °C for 22 h, and the hydrolyzed solutions were tested for Trp through a F-7000 fluorescence detector (Hitachi) with an excitation wavelength of 280 nm and an emission wavelength of 340 nm, as described by Gold et al. [30].

2.6. Statistical Analyses

All the statistical analyses were carried out using R 3.4.1 [33]. Differences among groups in the Trials 1 and 2 were tested using analysis of variance (ANOVA) provided by the *multcomp* package [34], which was associated with the *TukeyHSD* function for the pairwise comparison of means. Significance was defined as $p < 0.05$.

3. Results

3.1. Effects of Nitrogen Source on the Food Waste Treatment

Nitrogen source showed limited impacts on the larval length ($df = 9$, $F = 2.287$, $p = 0.059$) and crude protein content ($df = 9$, $F = 1.398$, $p = 0.253$), while in contrast, exerting significant effects on the larval weight ($df = 9$, $F = 8.167$, $p < 0.001$), protein yield ($df = 9$, $F = 16.410$, $p < 0.001$), crude lipid content ($df = 9$, $F = 3.800$, $p = 0.01$), and lipid yield ($df = 9$, $F = 20.690$, $p < 0.001$) as shown in the Figure 1. After 12 days of treatment, the larval body length reached 13.4–15.4 mm (Figure 1A) and crude protein content increased to 36.3–40.6%, regardless of the nitrogen source difference (Figure 1C). For the larval weight, the $NaNO_3$ group showed significantly lighter weight than the other groups, with an average at 0.0423 g when other groups' averages ranged from 0.0688 to 0.0959 g (Figure 1B). For the protein yield, both NH_4Cl (3.5%) and $NaNO_3$ (1.1%) groups were significantly lower than that of the blank control (7.8%), while the results from other groups (7.1–9.3%) were similar to that of the control (Figure 1D). For the crude lipid content, none of the nitrogen source groups were significantly different from the blank control (25.1%), ranging from 21.3% to 30.9% (Figure 1E). For the lipid yield, results from NH_4Cl (2.6%) and $NaNO_3$ (0.8%) groups were significantly lower than that of the blank control (4.9%), with the result of the soybean flour group (8.3%) significantly higher than that of the control and results from the other groups (4.7–6.9%) similar to that of the control (Figure 1F).

Nitrogen sources substantially affected the process efficiency, as shown by the indexes of WR ($df = 9$, $F = 90.640$, $p < 0.001$), LY ($df = 9$, $F = 25.080$, $p < 0.001$), SR ($df = 9$, $F = 15.52$, $p < 0.001$), NCR ($df = 9$, $F = 18.88$, $p < 0.001$), and ECD ($df = 9$, $F = 5.438$, $p = 0.001$), respectively in the Figure 2. The NH_4Cl (45.8%) and $NaNO_3$ (12.3%) groups displayed significantly lower WR index compared to the blank control (63.1%), whereas the urea group (73.5%) exhibited significantly higher WR than the control, while results from other groups (58.8–72.0%) were similar to that of the control (Figure 2A). The LY and SR indexes exhibited similar trends, that NH_4Cl (LY 8.6%, SR 42.0%) and $NaNO_3$ (LY 3.01%, SR 29.1%) groups were significantly lower than the blank control (LY 19.3%, SR 87.3%), and the other groups (LY 17.8–25.4%, SR 81.7–99.2%) were similar to the control (Figure 2B,C). The NCR index showed that results from the NH_4Cl (23.1%) and $NaNO_3$ (7.2%) groups were significantly lower than that of the control (73.5%), while data from the urea (50.0%), uric acid (48.9%), and Glu (46.8%) groups were higher than that of the NH_4Cl and $NaNO_3$ groups, whereas lower than the control, and data from all other groups (57.7–60.7%) were similar to that of the control (Figure 2D). The ECD indexes varied between 18.9% and 38.7% (Figure 2E), and none of the nitrogen groups demonstrated significantly different ECD compared to the control (30.6%). As for the mass balance analysis, the mass distribution pattern suggested that NH_4Cl and $NaNO_3$ groups resulted in substantially less larvae yield and higher frass residues compared to the blank control, while other nitrogen groups showed approximately similar distribution patterns of larvae, frass, and metabolism mass as the control (Figure 2F).

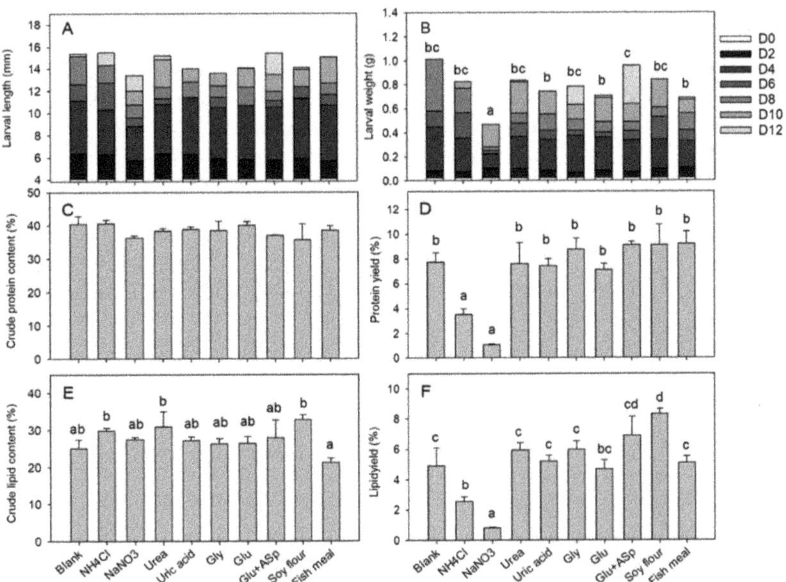

Figure 1. Larval growing and nutrient properties following nitrogen source amendment in Trial 1. Error bars represent standard deviations (*n* = 3). The (**A**) larval body length and (**C**) crude protein content show no significant differences among groups. The (**B**) larval body weight ($p < 0.001$), (**D**) protein yield ($p < 0.001$), (**E**) crude lipid content ($p = 0.010$), and (**F**) lipid yield ($p < 0.001$) are different among groups, and groups with different letters represent significant difference ($p < 0.05$).

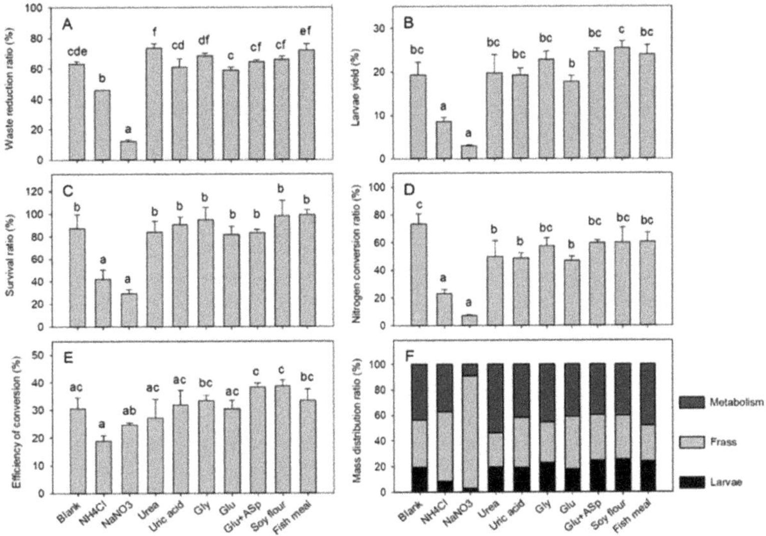

Figure 2. Bioconversion efficiencies following nitrogen source amendment in Trial 1. Error bars represent standard deviations (*n* = 3). The (**A**) waste reduction ratio ($p < 0.001$), (**B**) larvae yield ($p < 0.001$), (**C**) survival ratio ($p < 0.001$), (**D**) nitrogen conversion ratio ($p < 0.001$), and (**E**) efficiency of conversion ($p = 0.001$) are different among groups, and groups with different letters represent significant difference ($p < 0.05$). The (**F**) mass distribution ratio shows no significant differences between groups.

3.2. Effects of C/N on the Food Waste Treatment

The C/N of food waste did not affect larval length ($df = 5$, $F = 0.488$, $p = 0.779$); however, it did influence larval weight ($df = 5$, $F = 5.934$, $p = 0.005$), crude protein content ($df = 5$, $F = 4.962$, $p = 0.011$), protein yield ($df = 5$, $F = 7.107$, $p = 0.003$), crude lipid content ($df = 5$, $F = 9.874$, $p = 0.001$), and lipid yield ($df = 5$, $F = 8.035$, $p = 0.002$) as shown in the Figure 3. Within the 6 C/N groups, larval length ranged from 14.4 to 15.9 mm (Figure 3A). Larval weight varied between 0.074 and 0.109 g (Figure 3B), and the C/N (14:1) group (0.109 g) showed significantly higher body weight than the blank control (0.081 g). Larval crude protein ratio ranged from 35.4% to 42.0%, with none of the 5 C/N group differing from the blank control significantly (Figure 3C). Protein yield ranged from 7.3% to 11.2%, and the C/N (16:1) group (11.2%) yielded significantly higher values than that of the blank control (7.8%) (Figure 3D). Larval crude lipid content changed from 21.2% to 30.3%. None of the 5 C/N groups significantly differed from the blank control, though data from the C/N (18:1–16:1) groups were higher than that of the C/N (14:1–10:1) groups (Figure 3E). Lipid yield varied between 4.4% and 8.1%, with the C/N (18:1) (7.6%) and C/N (16:1) (8.1%) groups exhibiting significantly higher data than the blank control (4.9%) (Figure 3F) and other C/N groups (4.4–6.2%), demonstrating results similar to the control.

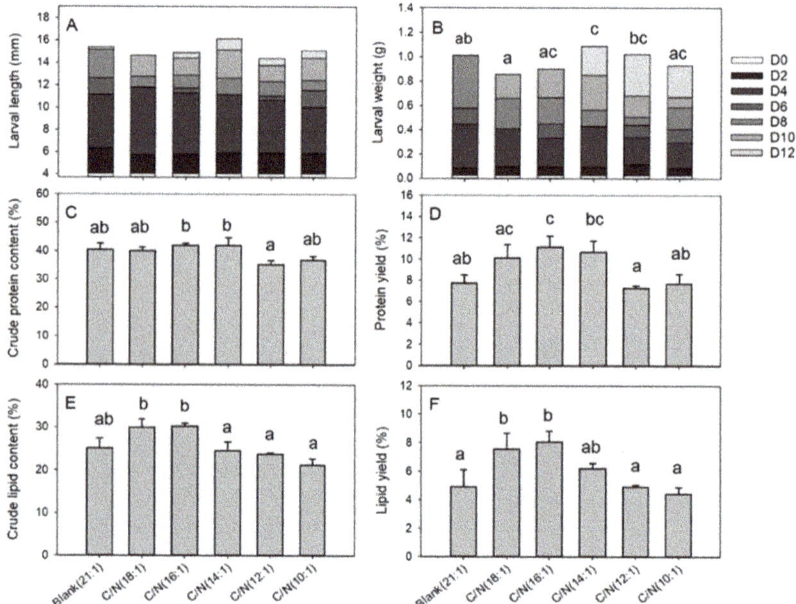

Figure 3. Larval growing and nutrient properties following C/N adjustment in Trial 2. Error bars represent standard deviations ($n = 3$). The (**A**) larval length shows no significant differences between groups. The (**B**) larval weight ($p = 0.005$), (**C**) crude protein content ($p = 0.011$), (**D**) protein yield ($p = 0.003$), (**E**) crude lipid content ($p = 0.001$), and (**F**) lipid yield ($p = 0.002$) are different among groups, and groups with different letters represent significant difference ($p < 0.05$).

The C/N of food waste greatly affected the process efficiency, as indicated by the WR ($df = 5$, $F = 29.630$, $p < 0.001$), LY ($df = 5$, $F = 4.540$, $p = 0.015$), SR ($df = 5$, $F = 6.468$, $p = 0.004$), and NCR ($df = 5$, $F = 19.400$, $p < 0.001$) indexes in the Figure 4. However, the ECD ($df = 5$, $F = 1.583$, $p = 0.238$) index was less affected. The WR indexes ranged from 61.0% to 84.8%, and the C/N (18:1) (84.8%), C/N (16:1) (77.6%), and C/N (14:1) (73.5%) groups yielded significantly higher data than that of the blank control (63.1%) (Figure 4A). The LY indexes changed from 19.3% to 26.6%, and the C/N (16:1) (26.6%) group was significantly higher than the blank control (19.3%), while the other 4 C/N groups ranged from 20.6% to 25.4%

of BCRs (Figure 4B). The SR indexes changed from 61.9% to 98.0%, and none of the 5 C/N groups differed from the blank control significantly (Figure 4C). The NCR indexes showed a pattern which indicated that the C/N (18:1) (83.0%), C/N (16:1) (81.3%), and C/N (14:1) (68.1%) groups were similar to the blank control (73.5%), while the C/N (12:1) (39.9%) and C/N (10:1) (35.2%) groups were lower than the control (Figure 4D). The ECD indexes varied within 29.7% and 34.6% and none of the 5 C/N groups differed significantly from the blank control (30.6%) (Figure 4E). The mass balance analysis suggested that the C/N (18:1), C/N (16:1), and C/N (14:1) groups together resulted in relatively higher larval ratios and lower frass ratios compared to the blank control, C/N (12:1), and C/N (10:1) groups (Figure 4F).

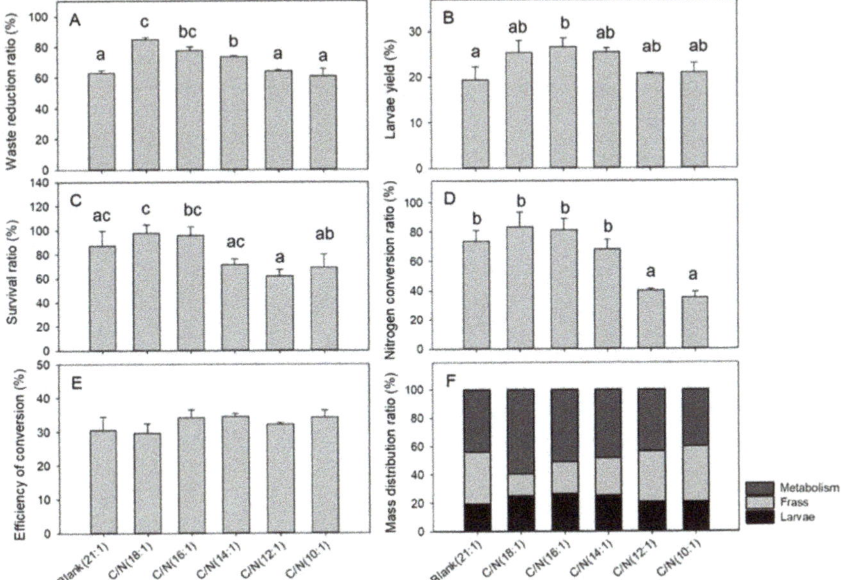

Figure 4. Bioconversion efficiencies following C/N adjustment in Trial 2. Error bars represent standard deviations ($n = 3$). The (**A**) waste reduction ratio ($p < 0.001$), (**B**) larvae yield ($p = 0.015$), (**C**) survival ratio ($p = 0.004$), and (**D**) nitrogen conversion ratio ($p < 0.001$) are different among groups, and groups with different letters represent significant difference ($p < 0.05$). The (**E**) efficiency of conversion and (**F**) mass distribution ratio show no significant differences among groups.

3.3. Effects of C/N on Larval Enzyme Activity and Amino Acid Composition

Among all the four larval enzymes tested, only the SOD activity ($df = 5$, $F = 4.561$, $p = 0.015$) differed greatly between the 6 C/N groups as shown in the Table 2, while the POD ($df = 5$, $F = 2.266$, $p = 0.114$), CAT ($df = 5$, $F = 1.710$, $p = 0.207$), and GSH-px ($df = 5$, $F = 0.816$, $p = 0.561$) activities did not. For the amino acid composition analysis, the compositions of each amino acid among the 6 C/N groups were not significantly different as shown in the Table 3. The averaged proportions of each amino acid were therefore calculated, and the Glu, Ala, and Asp were found to be the top 3 most abundant amino acids.

Table 2. Activities of antioxidant enzyme of larvae collected on Day 12 in Trial 2.

Groups	POD (U/g)	SOD (U/g)	CAT (U/g)	GSH-px (U/g)
Blank (21:1)	0.056 ± 0.079	0.149 ± 0.082 [ab]	0.071 ± 0.034	7.40 ± 0.78
C/N (18:1)	0.00 ± 0.00	0.249 ± 0.081 [b]	0.094 ± 0.026	8.46 ± 1.77
C/N (16:1)	0.049 ± 0.048	0.213 ± 0.133 [ab]	0.035 ± 0.034	9.91 ± 4.56
C/N (14:1)	0.00 ± 0.00	0.058 ± 0.009 [ab]	0.039 ± 0.027	7.12 ± 2.03
C/N (12:1)	0.170 ± 0.121	0.00 ± 0.00 [a]	0.037 ± 0.018	4.90 ± 0.62
C/N (10:1)	0.00 ± 0.00	0.00 ± 0.00 [a]	0.042 ± 0.008	5.94 ± 1.39
	$F = 2.266, p = 0.114$	$F = 4.561, p = 0.015$	$F = 1.710, p = 0.207$	$F = 0.816, p = 0.561$

Values are presented as mean $\pm$ standard deviation ($n = 3$). Different letters represent significant differences among column-wise groups.

Table 3. Proximate amino acid compositions of larvae collected on Day 12 in Trial 2.

Amino Acids	Blank (21:1)	C/N (18:1)	C/N (16:1)	C/N (14:1)	C/N (12:1)	C/N (10:1)	Overall
Cys	1.0 ± 0.1	0.9 ± 0.1	0.8 ± 0.0	0.9 ± 0.0	0.8 ± 0.0	0.8 ± 0.0	0.9 ± 0.1
Met	3.3 ± 0.2	3.4 ± 0.1	3.2 ± 0.1	3.3 ± 0.2	3.2 ± 0.2	3.1 ± 0.1	3.2 ± 0.2
Asp	8.5 ± 0.4	9.2 ± 1.0	9.1 ± 0.8	8.8 ± 0.3	9.4 ± 0.8	9.5 ± 0.6	9.1 ± 0.8
Thr	4.4 ± 0.2	4.5 ± 0.3	4.5 ± 0.2	4.3 ± 0.2	4.5 ± 0.2	4.5 ± 0.2	4.4 ± 0.2
Ser	5.0 ± 0.2	5.0 ± 0.3	5.1 ± 0.3	5.0 ± 0.2	4.9 ± 0.2	5.0 ± 0.2	5.0 ± 0.3
Glu	10.6 ± 0.5	11.2 ± 0.7	11.6 ± 0.3	11.7 ± 1.2	10.8 ± 0.6	12.3 ± 0.5	11.4 ± 0.9
Gly	6.2 ± 0.3	6.0 ± 0.2	6.2 ± 0.2	6.1 ± 0.3	5.9 ± 0.3	5.8 ± 0.2	6.0 ± 0.3
Ala	11.1 ± 0.3	10.2 ± 0.5	9.9 ± 0.6	10.0 ± 0.8	10.0 ± 0.2	9.1 ± 0.3	10.0 ± 0.8
Val	7.0 ± 0.5	6.9 ± 0.5	7.0 ± 0.3	6.7 ± 0.2	6.8 ± 0.5	6.4 ± 0.4	6.8 ± 0.4
Ile	4.9 ± 0.3	4.8 ± 0.3	4.8 ± 0.2	4.8 ± 0.2	4.9 ± 0.3	4.8 ± 0.3	4.8 ± 0.3
Leu	7.5 ± 0.2	7.5 ± 0.5	7.5 ± 0.3	7.5 ± 0.1	7.9 ± 0.4	7.8 ± 0.1	7.6 ± 0.3
Tyr	0.7 ± 0.1	0.6 ± 0.1	0.5 ± 0.2	0.5 ± 0.0	0.6 ± 0.1	0.5 ± 0.0	0.6 ± 0.1
Phe	4.5 ± 0.5	4.7 ± 0.4	4.6 ± 0.1	4.4 ± 0.2	4.7 ± 0.5	4.6 ± 0.3	4.6 ± 0.4
Lys	5.9 ± 0.2	5.9 ± 0.1	6.1 ± 0.2	6.0 ± 0.2	6.3 ± 0.1	6.4 ± 0.2	6.1 ± 0.3
His	3.5 ± 0.2	3.6 ± 0.2	3.5 ± 0.1	3.5 ± 0.1	3.3 ± 0.2	3.1 ± 0.1	3.4 ± 0.2
Arg	5.2 ± 0.1	4.9 ± 0.5	4.8 ± 0.4	5.2 ± 0.0	5.0 ± 0.1	5.3 ± 0.1	5.1 ± 0.3
Pro	7.3 ± 0.1	7.1 ± 0.3	7.0 ± 1.1	7.5 ± 0.4	7.2 ± 0.1	7.5 ± 0.4	7.3 ± 0.5
Trp	1.3 ± 0.4	1.6 ± 0.4	1.7 ± 0.1	1.6 ± 0.4	1.9 ± 0.1	1.5 ± 0.0	1.6 ± 0.3

Values are presented as g/100 g protein with mean $\pm$ standard deviation ($n = 3$).

4. Discussion

4.1. Effects of Nitrogen Source on the Bioconversion Process

Among all the 9 nitrogen sources, the NH_4Cl and $NaNO_3$ resulted in markedly adverse effects on the larval development and process efficiency compared to the 7 other organic nitrogen species, suggesting that NH_4Cl and $NaNO_3$ were probably less suitable than the organic nitrogen species in terms of facilitating larval development and waste degradation. The less efficient performance in the NH_4Cl and $NaNO_3$ conditions were probably due to several reasons, including: (1) the high amount of compounds used, (2) the toxicity generated by chloride or sodium salt, and (3) the low survival and adaptability of BSFL to these environments. In general, NH_4-N and NO_3-N were typical nitrogen species in the leachate and sludge biowaste. Larvae that fed on liquid leachate [24] and sewage sludge [18] have been found to be of low survival rate (30–40% mortality) and low bioconversion performance (LY 0.2–2.3%), which complied with the present results in the NH_4Cl and $NaNO_3$ conditions. However, Green et al. [23] reported that feeding BSFL with 10 mM $NaNO_3$ solution, specifically 14 mg NO_3-N/100 mL solution, facilitated the BSFL's transformation of NO_3-N to NO_2-N and further to NH_4-N (e.g., denitrification). The current study supplied 1 g NH_4-N or NO_3-N/100 g dry matter to the food waste. The high amount of N element could be one of the main reasons for the negative effects of NH_4Cl and $NaNO_3$ effects on the BSFL compared to Green's study. However, as a nitrogen source used for nitrogen amendment of food waste, addition of 1 g N/100 g dry matter is a reasonable requirement that NH_4-N and NO_3-N might fail to address the need due to their negative effects on BSFL.

Interestingly, a subpopulation of larvae survived after the NH_4Cl and $NaNO_3$ amendments in the study. This subpopulation resulted in a similar protein and lipid body ratio when compared with the larvae which grew in the 7 other organic nitrogen conditions, suggesting that a small percentage of BSFL may gain the ability to adapt to NH_4Cl and $NaNO_3$ nutrient/environment, either through the direct incorporation of the NH_4-N/NO_3-N or through the indirect utilization of NH_4-N/NO_3-N assimilated by the in vivo or in vitro microorganisms. Barragan-Fonseca et al. [35] has pointed out that the larval protein content is regulated within narrow constraints, whereas the fat content is strongly impacted by nutrient concentration. The present results agreed with these findings that larval protein content was limitedly affected by the nitrogen environment in terms of BSFL survival; however, the larval lipid content was significantly affected by the nitrogen sources, where urea, L-Glu/L-Asp, and soybean flour were probably better nutrient sources for BSFL compared to other nitrogen species due to the relatively higher lipid content or lipid yield of BSFL.

Food waste amended with the 7 organic nitrogen sources generally resulted in neither negative nor positive effects on the BSFL performance compared to the control based on the bioconversion indexes of WR, LY, SR, and ECD. One of the possible reasons could be the inefficient supplemental ratio of the organic nitrogen sources to the food waste. According to the NCR index, all the nitrogen amendment conditions were lower than the control, suggesting that the optimum nitrogen supplying amount was not achieved, and the NCR values (46.8–60.7%) of the 7 organic nitrogen conditions were less than a previous study (66.4% $\pm$ 6.5%) conducted for mussel and bread waste treatment [13]. These results suggest that the addition amount of selected nitrogen source is very important and should be critically optimized in order to achieve high performance of waste reduction and BSFL bioconversion efficiency.

4.2. Effects of C/N on the Bioconversion Process

Although there is no well-known regulation about whether urea could be used as a food additive for the BSFL, urea is a suitable nitrogen source for investigating the C/N effects on the food waste treatment by the BSFL. The reasons are as follow: (1) Urea is a nitrogen source containing the highest N content (46.7%) and lowest C content (20.0%) amongst the 7 organic nitrogen species, which allow urea to be one of the most efficient nitrogen sources used for nitrogen content amendment while simultaneously limiting the energy/nutrient effect generated by the carbon. (2) Urea exhibited a feasible nutrient effect to BSFL according to the lipid and protein analysis in the Trial 1, (3) urea is a natural food source for BSFL as it is contained in the animal feces [19,21], and (4) urea is widely available as an artificially synthesized chemical. Therefore, urea was selected as the nitrogen source used for C/N adjustment in the Trial 2 of this study.

The larval protein and lipid data suggested that the C/N range of 18:1 to 16:1 was optimal for a high larval protein and lipid yield, while the bioconversion indexes indicated that the C/N range of 18:1 to 14:1 was highly efficient for the waste reduction (WR, 73.5–84.8%, $p < 0.001$) and larvae production (LY, 10.1–11.1%, $p = 0.003$). Therefore, supplying food waste with a moderate amount of urea (adjusting food waste C/N from 21:1 down to 18:1–14:1, especially 18:1–16:1) significantly facilitated larval development and food waste consumption; however, further addition of urea (lowering the C/N down to 12:1–10:1) would result in urea waste and even negative effects on the larval growing and bioconversion process, as indicated by the declining WR, LY, and NCR indexes.

The current results suggest that overdosing nitrogen-rich material would result in larval mortality and declined process efficiency. Similar results have also been seen in two other recent studies. Lopes et al. [14] studied recycling aquaculture waste by feeding BSFL with fish waste and bread mixture, and Ewald et al. [13] has tried to manipulate larval fatty acid composition by feeding BSFL with mussel and bread mixture. Both studies suggested that adding a moderate amount of nitrogen-rich aquaculture waste was beneficial for larval development, while too much nitrogen material may lead to negative effects such as larval mortality and biomass loss. Based on the larvae yield, this moderate range for fish waste treatment was approximately 5–15% of fish carcasses [14], and approximately 10–20% of mussel for the mussel material [13]. Unfortunately, neither studies reported the C or N content of the diet materials. If roughly assuming the bread C and N content [36] to be 48.9% of C (50% of organic matter) and 2.95% of N (16% of crude protein), and fish C and N content to be 45% of C and 66% of N (as fish meal in this study), the 5–15% of fish waste in Lopes' study is thus equivalent to C/N of 15:1–12:1, which is close to the optimum range of 18:1–14:1 found in this study, indicating the beneficial effects of modifying C/N of food waste into this range. As for the applying amount of nitrogen amendment, urea could be more efficient than aquaculture waste, as 0.36–1.10 g urea/100 g dry matter (equivalent to 0.11–0.33% wet weight basis) used in this study resulted in comparable performance improvements compared with the 5–15% of fish waste [14] or 10–20% of mussel waste [13] used in previous studies.

Interestingly, the C/N adjustment through urea altered the larval production performance but not the crude protein content or the amino acid composition. In the fish and mussel studies [13,14], a clear trend of the positive correlations between the aquaculture materials and larval protein content was observed, although higher larval mortality occurred simultaneously with more aquaculture waste used. This finding suggests that the urea may not support the BSFL growth directly with the amino acid nutrient, whereas it may improve larval development through modulating larval metabolism, such as the SOD enzyme activity. Interestingly, the amino acid composition of BSFL is relatively stable despite wide variations of C/Ns. This could be highly due to the same nitrogen source, i.e., urea, used in the current C/N trial. In another study where BSFL fed on different substrates [18], the amino acid composition among groups varied greatly, and Tyr, Glu, and Asp were found as the top 3 amino acids for BSFL fed on human feces, and Glu, Asp, and Lys were found as the top 3 species for the food waste substrates. The current study also found Glu and Asp as the top 2 amino acids, whereas the third abundant species was Ala. Compared to the previous study [18], the Cys, Met, Thr, Ser, Gly, and Pro proportions were generally higher, and the Tyr and Lys proportions were generally lower in the larvae of current study. These results suggest that the nitrogen species may influence BSFL amino acid composition more greatly than the C/N ratios.

Altogether, the waste reduction performances of 73.5–84.8% and larvae yield of 25.3–26.6% at the C/N conditions of 18:1–14:1 in the current study are higher than many of the previous BSFL studies [3,19]. Other than the higher nutrient and digestibility of food waste used in current study, the nitrogen supplement of urea and optimal range of C/N could be two of the main contributors to the performance improvement.

5. Conclusions

Adjusting the C/N of food waste substrate is a viable method for improving the larval treatment performance. Organic nitrogen is more suitable than the NH_4Cl or $NaNO_3$ as the nitrogen amendment. Urea was a reliable and practical nitrogen source for the C/N adjustment. Addition of small amounts of urea, targeting C/N of 18:1–14:1, may significantly improve the waste reduction performance, while targeting C/N of 18:1–16:1 may substantially increase the larval protein and lipid conversion efficiency, and the BSFL amino acid composition was not affected by the C/N variation. Therefore, the current study reveals that the C/N range of 18:1–16:1 is likely the optimal condition for food waste treatment by BSFL, and the application of the current strategy may improve the food waste biodegradation and facilitate the nutrient recycling by the BSFL farming.

Author Contributions: Conceptualization, W.X., H.Z. and J.X.; methodology, W.X. and M.W.; validation, Y.B. and H.T.; investigation, Y.L., S.Z. and S.S.; data curation, Y.L., N.J. and M.R.; writing—original draft preparation, Y.L., S.Z. and W.X.; writing—review and editing, J.X., H.Z. and W.X.; funding acquisition, W.X. All authors have read and agreed to the published version of the manuscript.

Funding: This study was financially supported by the Fundamental Research Funds for the Central Universities (grant number DUT21LK29, DUT20LK36), Yingkou enterprise and doctor innovation program of Yingkou Science and Technology Bureau (grant number 202005), and the National Natural Science Foundation of China (grant numbers 31670767).

Institutional Review Board Statement: Not applicable.

Data Availability Statement: The data presented in this study are openly available in FigShare at doi.10.6084/m9.figshare.14609535.

Conflicts of Interest: The authors declare no conflict of interest.

References

1. Lee, K.; Yun, E.; Goo, T. Antimicrobial Activity of an Extract of *Hermetia illucens* Larvae Immunized with Lactobacillus casei against Salmonella Species. *Insects* **2020**, *11*, 704. [CrossRef]
2. Fischer, H.; Romano, N.; Sinha, A.K. Conversion of Spent Coffee and Donuts by Black Soldier Fly (*Hermetia illucens*) Larvae into Potential Resources for Animal and Plant Farming. *Insects* **2021**, *12*, 332. [CrossRef]
3. Cickova, H.; Newton, G.L.; Lacy, R.C.; Kozanek, M. The use of fly larvae for organic waste treatment. *Waste Manag.* **2015**, *35*, 68–80. [CrossRef] [PubMed]
4. Tomberlin, J.K.; Huis, A.V. Black soldier fly from pest to 'crown jewel' of the insects as feed industry: An historical perspective. *J. Insects Food Feed.* **2020**, *6*, 1–4. [CrossRef]
5. Zhang, J.B.; Tomberlin, J.K.; Cai, M.M.; Xiao, X.P.; Zheng, L.Y.; Yu, Z.N. Research and industrialisation of *Hermetia illucens* L. in China. *J. Insects Food Feed.* **2020**, *6*, 5–12. [CrossRef]
6. Gladun, V.V. The first record of *Hermetia illucens* (Diptera, Stratiomyidae) from Russia. *Nat. Conserv. Res.* **2019**, *4*. [CrossRef]
7. Marshall, S.A.; Woodley, N.E.; Hauser, M. The historical spread of the Black Soldier Fly, *Hermetia illucens* (L.) (Diptera, Stratiomyidae, Hermetiinae), and its establishment in Canada. *J. Kans. Entomol. Soc.* **2015**, *146*, 51–56.
8. Üstüner, T.; Hasbenlí, A.; Rozkošný, R. The first record of *Hermetia illucens* (Linnaeus, 1758) (Diptera, Stratiomyidae) from the Near East. *Stud. Dipterol.* **2003**, *10*, 181–185. [CrossRef]
9. Holmes, L.A.; Vanlaerhoven, S.L.; Tomberlin, J.K. Lower temperature threshold of black soldier fly (Diptera: Stratiomyidae) development. *J. Insects Food Feed.* **2016**, *2*, 1–8. [CrossRef]
10. Harnden, L.M.; Tomberlin, J.K. Effects of temperature and diet on black soldier fly, *Hermetia illucens* (L.) (Diptera: Stratiomyidae), development. *Forensic Sci. Int.* **2016**, *266*, 109–116. [CrossRef]
11. Cammack, J.; Tomberlin, J. The Impact of Diet Protein and Carbohydrate on Select Life-History Traits of The Black Soldier Fly *Hermetia illucens* (L.) (Diptera: Stratiomyidae). *Insects* **2017**, *8*, 56. [CrossRef]
12. Beesigamukama, D.; Mochoge, B.; Korir, N.K.; Fiaboe, K.; Tanga, C.M. Biochar and gypsum amendment of agro-industrial waste for enhanced black soldier fly larval biomass and quality frass fertilizer. *PLoS ONE* **2020**, *15*, e238154. [CrossRef]
13. Ewald, N.; Vidakovic, A.; Langeland, M.; Kiessling, A.; Sampels, S.; Lalander, C. Fatty acid composition of black soldier fly larvae (*Hermetia illucens*)—Possibilities and limitations for modification through diet. *Waste Manag.* **2020**, *102*, 40–47. [CrossRef]
14. Lopes, I.G.; Lalander, C.; Vidotti, R.M.; Vinnerås, B. Using *Hermetia illucens* larvae to process biowaste from aquaculture production. *J. Clean. Prod.* **2020**, *251*, 119753. [CrossRef]
15. Pang, W.; Hou, D.; Nowar, E.E.; Chen, H.; Wang, S. The influence on carbon, nitrogen recycling, and greenhouse gas emissions under different C/N ratios by black soldier fly. *Environ. Sci. Pollut. Res.* **2020**, *27*, 42767–42777. [CrossRef]
16. Li, Q.; Zheng, L.; Hao, C.; Garza, E.; Zhou, S. From organic waste to biodiesel: Black soldier fly, *Hermetia illucens*, makes it feasible. *Fuel* **2011**, *90*, 1545–1548. [CrossRef]
17. Gold, M.; Tomberlin, J.K.; Diener, S.; Zurbrügg, C.; Mathys, A. Decomposition of biowaste macronutrients, microbes, and chemicals in black soldier fly larval treatment: A review. *Waste Manag.* **2018**, *82*, 302–318. [CrossRef] [PubMed]
18. Lalander, C.; Vinnerås, B.; Sveriges, L. Effects of feedstock on larval development and process efficiency in waste treatment with black soldier fly (*Hermetia illucens*). *J. Clean. Prod.* **2019**, *208*, 211–219. [CrossRef]
19. Surendra, K.C.; Tomberlin, J.K.; van Huis, A.; Cammack, J.A.; Heckmann, L.L.; Khanal, S.K. Rethinking organic wastes bioconversion: Evaluating the potential of the black soldier fly (*Hermetia illucens* (L.)) (Diptera: Stratiomyidae) (BSF). *Waste Manag.* **2020**, *117*, 58–80. [CrossRef] [PubMed]
20. Oonincx, D.G.A.B.; Broekhoven, V.S.; Huis, V.A.; Loon, V.J.J.A. Feed Conversion, Survival and Development, and Composition of Four Insect Species on Diets Composed of Food By-Products. *PLoS ONE* **2015**, *10*, e144601. [CrossRef] [PubMed]
21. Rehman, K.U.; Cai, M.; Xiao, X.; Zheng, L.; Hui, W.; Soomro, A.A.; Zhou, Y.; Wu, L.; Yu, Z.; Zhang, J. Cellulose decomposition and larval biomass production from the co-digestion of dairy manure and chicken manure by mini-livestock (*Hermetia illucens* L.). *J. Environ. Manag.* **2017**, *196*, 458–465. [CrossRef]
22. Fischer, H.; Romano, N. Fruit, vegetable, and starch mixtures on the nutritional quality of black soldier fly (*Hermetia illucens*) larvae and resulting frass. *J. Insects Food Feed.* **2021**, *7*, 319–327. [CrossRef]
23. Green, T.R.; Popa, R. Enhanced Ammonia Content in Compost Leachate Processed by Black Soldier Fly Larvae. *Appl. Biochem. Biotechnol.* **2012**, *166*, 1381–1387. [CrossRef]

24. Grossule, V.; Lavagnolo, M.C. The treatment of leachate using Black Soldier Fly (BSF) larvae: Adaptability and resource recovery testing. *J. Environ. Manag.* **2020**, *253*, 109707. [CrossRef] [PubMed]
25. Liu, X.; Chen, X.; Wang, H.; Yang, Q.; Rehman, K.U.; Li, W.; Cai, M.; Li, Q.; Mazza, L.; Zhang, J.; et al. Dynamic changes of nutrient composition throughout the entire life cycle of black soldier fly. *PLoS ONE* **2017**, *12*, e0182601. [CrossRef]
26. Erickson, M.C.; Islam, M.; Sheppard, C.; Liao, J.; Doyle, M.P. Reduction of Escherichia coli O157:H7 and Salmonella enterica Serovar Enteritidis in Chicken Manure by Larvae of the Black Soldier Fly. *J. Food Prot.* **2004**, *67*, 685. [CrossRef]
27. Liu, Y.; Wang, W.; Xu, J.; Xue, H.; Stanford, K.; McAllister, T.A.; Xu, W.; Martinez-Abarca, F. Evaluation of compost, vegetable and foozzd waste as amendments to improve the composting of NaOH/NaClO-contaminated poultry manure. *PLoS ONE* **2018**, *13*, e205112. [CrossRef]
28. Janssen, R.H.; Vincken, J.; van den Broek, L.A.M.; Fogliano, V.; Lakemond, C.M.M. Nitrogen-to-Protein Conversion Factors for Three Edible Insects: Tenebrio molitor, Alphitobius diaperinus, and *Hermetia illucens*. *J. Agric. Food Chem.* **2017**, *65*, 2275–2278. [CrossRef] [PubMed]
29. Zheng, L.; Li, Q.; Zhang, J.; Yu, Z. Double the biodiesel yield: Rearing black soldier fly larvae, *Hermetia illucens*, on solid residual fraction of restaurant waste after grease extraction for biodiesel production. *Renew. Energy* **2012**, *41*, 75–79. [CrossRef]
30. Gold, M.; Cassar, C.M.; Zurbrügg, C.; Kreuzer, M.; Boulos, S.; Diener, S.; Mathys, A. Biowaste treatment with black soldier fly larvae: Increasing performance through the formulation of biowastes based on protein and carbohydrates. *Waste Manag.* **2020**, *102*, 319–329. [CrossRef] [PubMed]
31. Diener, S.; Zurbrügg, C.; Tockner, K. Conversion of organic material by black soldier fly larvae: Establishing optimal feeding rates. *Waste Manag. Res.* **2009**, *27*, 603–610. [CrossRef] [PubMed]
32. Chen, E.; Wei, D.; Wei, D.; Yuan, G.; Wang, J. The effect of dietary restriction on longevity, fecundity, and antioxidant responses in the oriental fruit fly, Bactrocera dorsalis (Hendel) (Diptera: Tephritidae). *J. Insect Physiol.* **2013**, *59*, 1008–1016. [CrossRef]
33. R: The R Project for Statistical Computing. Available online: https://www.r-project.org/ (accessed on 14 April 2021).
34. Hothorn, T.; Bretz, F.; Westfall, P. Simultaneous Inference in General Parametric Models. *Biom. J.* **2008**, *50*, 346–363. [CrossRef]
35. Barragan-Fonseca, K.B.; Dicke, M.; Loon, V.J.J.A. Influence of larval density and dietary nutrient concentration on performance, body protein, and fat contents of black soldier fly larvae (Hermetia illucens). *Entomol. Exp. Appl.* **2018**, *166*, 761–770. [CrossRef] [PubMed]
36. Spranghers, T.; Ottoboni, M.; Klootwijk, C.; Ovyn, A.; Smet, S.D. Nutritional composition of black soldier fly (*Hermetia illucens*) prepupae reared on different organic waste substrates. *J. Sci. Food Agric.* **2017**, *97*, 2594–2600. [CrossRef]

Article

Self-Selection of Agricultural By-Products and Food Ingredients by *Tenebrio molitor* (Coleoptera: Tenebrionidae) and Impact on Food Utilization and Nutrient Intake

Juan A. Morales-Ramos [1,*], M. Guadalupe Rojas [1], Hans C. Kelstrup [2] and Virginia Emery [2]

1   Biological Control of Pests Research Unit, National Biological Control Laboratory, USDA-ARS, Stoneville, MS 38776, USA; guadalupe.rojas@usda.gov
2   Beta Hatch Inc., 200 Titchenal Road, Cashmere, WA 98815, USA; hans@betahatch.com (H.C.K.); virginia@betahatch.com (V.E.)
*   Correspondence: juan.moralesramos@usda.gov; Tel.: +1-662-686-3069

Received: 15 October 2020; Accepted: 20 November 2020; Published: 24 November 2020

Simple Summary: Insects have been considered as an alternative to fishmeal in animal feed formulations. Current methods for mass producing them remain expensive and, although cost is not the current market driver for insect products, they remain off the main stream. One way to reduce production costs is to lower the cost of insect diets. This could be accomplished by using agricultural by-products as ingredients to formulate insect diets. In this study 20 ingredients were tested as dietary components for the yellow mealworm. Ingredients included dry potato and cabbage; the bran of wheat and rice; by-product meals from vegetable oil production; spent distiller's grains from brewery and ethanol production; and hulls of different grains. A method called self-selection was used to approach the optimal proportion of these ingredients in mealworm diets by measuring their relative consumption. Nine combinations of eight ingredients were presented to groups of mealworms while carefully measuring the relative consumption of each ingredient. Results showed that the most suitable ingredients for mealworm production were dry cabbage and potato, the bran of wheat and rice, the meals of canola and sunflower, and distilled grains from corn and barley. This information will be used to formulate and evaluate diet formulations for the yellow mealworm in future research.

Abstract: Nutrient self-selection was used to determine optimal intake ratios of macro-nutrients by *Tenebrio molitor* L. larvae. Self-selection experiments consisted of 9 combinations (treatments) of 8 ingredients, from a total of 20 choices, radially distributed in a multiple-choice arena presented to groups of 100 *T. molitor* larvae (12th–13th instar). Larvae freely selected and feed on the pelletized ingredients for a period of 21 days at 27 °C, 75% RH, and dark conditions. Consumption (g) of each ingredient, larval live weight gained (mg), and frass production were recorded and used to calculate food assimilation and efficiency of conversion of ingested food. The macro-nutrient intake ratios were 0.06 ± 0.03, 0.23 ± 0.01, and 0.71 ± 0.03 for lipid, protein, and carbohydrate, respectively on the best performing treatments. The intake of neutral detergent fiber negatively impacted food assimilation, food conversion and biomass gain. Food assimilation, food conversion, and biomass gain were significantly impacted by the intake of carbohydrate in a positive way. Cabbage, potato, wheat bran, rice bran (whole and defatted), corn dry distillers' grain, spent brewery dry grain, canola meal and sunflower meal were considered suitable as *T. molitor* diets ingredients based on their relative consumption percentages (over 10%) within treatment.

Keywords: insects as feed and food; nutrition; food assimilation; food conversion; insect dietetics; insect rearing; macro-nutrients

---

1. Introduction

In recent years, the yellow mealworm, *Tenebrio molitor* L. (Coleoptera: Tenebrionidae) has been considered as a potential source of animal protein in feeds for fish [1–6] and livestock [7–14]. An increasing number of companies have been founded every year since 2013 that focus on insect mass production as animal feed [15]. One of the most important aspects of mass production involves the formulation of inexpensive yet effective diets that maximize biomass productivity over time. Recent research has focused on the potential use of agricultural by-products as insect food to reduce production costs of insect biomass [16–19]. However, current studies on *T. molitor* have focused on the effects of single ingredients on biological and food utilization parameters. No attempts have been made to evaluate combinations of multiple by-products as ingredients with the aim to develop diets for *T. molitor*. Self-selection studies incorporating by-products have been used to develop complete diets for the house cricket, *Acheta domesticus* L. [20].

Developing adequate diet formulations for insects using multiple undefined (oligidic) ingredients is a complex procedure and can take multiple years of research, particularly in insect species with long life cycles such as *T. molitor*. Optimal diets can be obtained from multiple oligidic ingredients by allowing insects to select the optimal ratios of each ingredient in a multiple-choice experimental setting. This method is known as self-selection and was first proposed by Waldbauer and Friedman (1991) [21]. The objectives of this study were to (1) determine the ingredients with the highest potential for formulation of insect diets using the self-selection method, (2) establish the optimal macro-nutrient intake ratios of *T. molitor* based on the self-selected intake of 20 ingredients, and (3) explore the impact of intake of macro-nutrients, neutral detergent fiber (NDF), phytosterol, and minerals including Fe, Mg, Ca, Zn, Cu, and Mn on the biomass gain, food assimilation and efficiency of conversion of ingested food (ECI).

2. Materials and Methods

2.1. Experimental Design

The colony stock, the rearing procedures and rearing hardware used in this study were as described by Morales-Ramos et al. [22]. Larvae of *T. molitor* used in the experiments were separated by size from the stock colony using sifters of standard numbers 10 and 12, which selected larvae with head capsule width measuring between 1.4 to 2 mm. According to estimates by Morales-Ramos et al. [23] these head capsule measurements correspond to larvae between 3 and 5 instars prior to pupation, which could include instars 11 to 14. However, because the stock used in this study has been selected for larger size, the experimental group of larvae could have included earlier instars.

Experimental units consisted of groups of 100 larvae maintained in multiple-choice arenas designed to provide equal access to 8 different food choices. Groups of larvae from each experimental unit were weighed at the beginning and end of the experiment and their weight was recorded for each of the units of each treatment. The multiple-choice arenas consisted of breathable round plastic dishes (120 × 25 mm, Pioneer Plastics 53C, Pioneer Plastics, North Dixon, KY, USA) modified by the addition of 8 sample plastic vials 20 mL (72 mm height × 25 mm diameter) (Product # 73400, Kartell s. p. a., Noviglio, Milan, Italy). The sample vials were cut to a height of 35 mm to fit inside the dishes and assembled perpendicularly to the dish in a radial pattern (45° apart) with equal distance to the center of the dish (Figure 1A). A 5 mm diameter opening was drilled into one side of each of the vials pointing perpendicularly to the center of the dish, to allow larvae to enter the vials (Figure 1B, a). A depression (90 × 2 mm) was constructed at the center of the dish with screened bottom (0.5 mm screen openings) to allow the collection of frass in a second dish located under the arena (Figure 1).

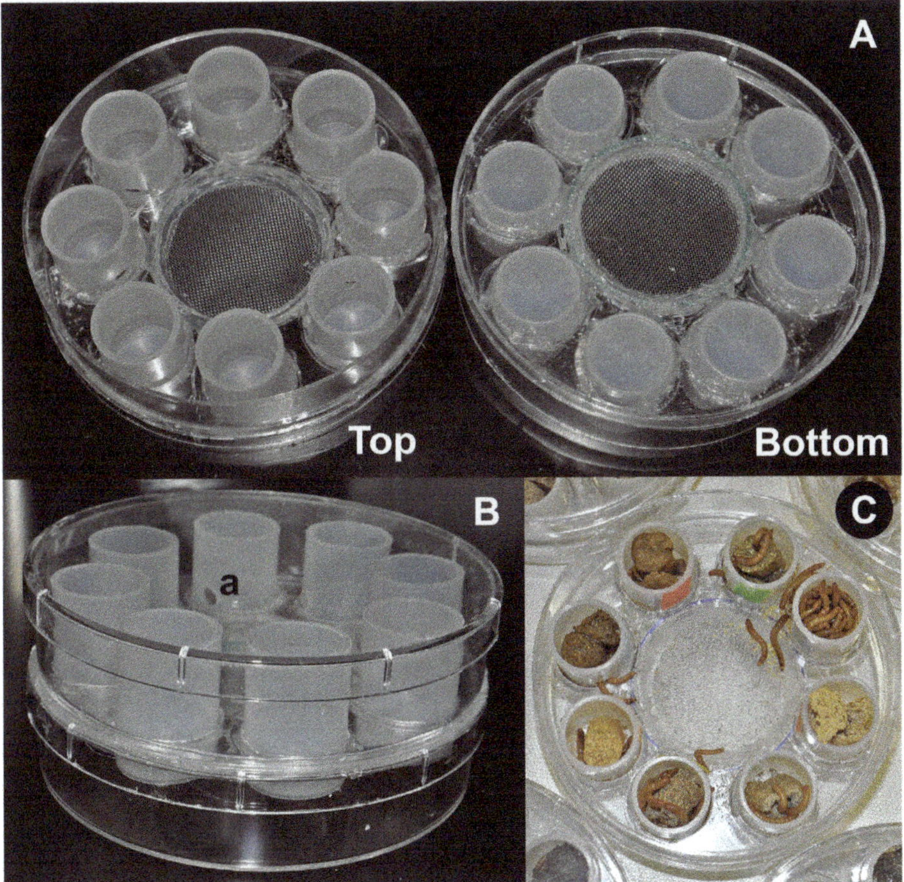

Figure 1. Multiple-choice arenas for self-selection experiments: (**A**) top and bottom views; (**B**) assembled arena with cover and bottom dish; (**C**) experimental unit with food choices and larvae, (a) opening into modified vial as food-choice compartment.

2.2. Food-Choice Treatments

Food choices consisted of 20 food products and agricultural by-products, which included dry white cabbage, potato flour, alfalfa pellets, wheat bran, rice bran (whole and solvent defatted), spelt screenings; meals from canola, soybean, olive, sunflower, cotton, and kelp; hulls from rice, oat, and peanut, and coffee chaff; and dry distilled grains from corn, wheat, and barley from ethanol production and brewery. Nine treatments of eight different combinations of these food choices were selected for the study (Table 1). The criterium of selection for the combination treatments was based on the relative content of each of the macro-nutrients (protein, lipid, and carbohydrate) and dietary fiber as neutral detergent fiber (NDF). Combinations should contain at least one food choice with high content of each of the macro-nutrients and dietary fiber to allow the mealworm larvae to select a complete diet. For instance, diet 1 contains canola meal as high protein ingredient, potato flour as high carbohydrate, corn distilled grain and dry cabbage for lipids and hulls from peanuts and rice as high fiber ingredients (Table 1).

Table 1. Combination of eight food choices presented to *Tenebrio molitor* larvae groups in nine self-selection treatments.

Ingredient	1	2	3	4	5	6	7	8	9
White Cabbage	X			X			X		X
Potato	X				X			X	
Alfalfa pellets	X	X						X	X
Wheat Bran	X		X			X			X
Rice bran whole			X	X	X	X	X		X
Rice bran defatted		X		X	X	X	X		
Spelt Screenings			X						
Corn dry distiller's grain	X	X			X	X	X		
Wheat dry distiller's grain				X		X			X
Barley brewery spent grain		X	X	X	X	X	X	X	
Canola meal	X	X		X	X		X		X
Soy Meal								X	X
Sunflower meal					X			X	
Olive meal				X					
Cotton seed meal			X						
Kelp meal			X						
Oat hulls			X			X	X	X	X
Peanut hulls	X	X	X	X	X	X	X	X	
Rice hulls	X	X							
Coffee chaff		X						X	

Food ingredients were ground into a fine powder using a high-speed food processor. Powdered food ingredients were individually mixed with reverse osmosis (RO) water at 50% to 70% ratio to obtain a consistency of dough. The food ingredients were then formed into sticks using a cut 10 mL syringe. These sticks were dried in a vacuum oven at 50 °C for a period of 48 h. This procedure resulted in stable dry sticks of each of the food ingredients listed in Table 1 with dimensions that allowed them to be introduced in the compartments of the multiple-choice arenas (Figure 1C).

Food combination treatments consisted of 10 repetitions each (=10 experimental units totaling 1000 larvae). Food ingredients were randomly distributed in the arena compartments to minimize the proximity effects among the different ingredients. At the beginning of the experiment, a measured amount of each of the corresponding food ingredients was added to the corresponding arena compartment in each of the experimental units. The initial amount of each food ingredient provided was recorded for each of the experimental units from each of the combination treatments. Experimental units were maintained in environmental chambers at 27 °C, 75% RH (relative humidity) and dark conditions for a period of three weeks. Experimental units were monitored daily to observe the consumption of each of the ingredients. Food ingredients that were depleted by consumption, were replenished with a measured amount of the corresponding food ingredient, which was recorded for each of the experimental units.

2.3. Data Collection and Analysis

At the end of a three-week period, larvae from each experimental unit were counted and weighed alive as a group. The remaining food was collected separately by ingredient, separated from frass, dried in a vacuum oven, and weighed. Frass was separated from food by sifting the remains using a standard No. 35 sieve (0.5 mm openings). The frass was collected, dried, and weighed using the same drying procedure. To collect the remaining food, all the vials in the arena were capped, the arena was inverted, and the contents of each vial were emptied, one by one, into a standard number 35 sieve by removing the cap to separate food from frass. The consumption of each ingredient (I_i) was calculated as total weight added of ingredient 'i'—remaining weight of ingredient 'i', were i = 1 to 8. The total food consumption (FC) was calculated as the sum of the consumption of all the eight ingredients.

Assimilated food (AF) was calculated as AF = FC − frass weight. The percent consumption of each ingredient was calculated as (I_i/FC) × 100. The weight of live mealworm biomass gained (LWG) was calculated as ending group weight—initial group weight. Mortality was extremely low (0.32 ± 0.21%) and dead larvae were cannibalized by surviving larvae (no cadavers were found), therefore, the ending live biomass measure per group included the loss of biomass due to mortality. Because the initial biomass dry weight could not be directly determined, the dry weight biomass gained (DWG) was calculated as LWG × the proportion of dry matter of mealworm larvae. The proportion of dry weight of mealworm larvae was previously determined from 25 groups of 10 larvae, which were weighed live, then frozen at −25 °C, dried in a vacuum oven at 50 °C, and weighed dry. The dry weight proportion of *T. molitor* late instar larvae was 0.38. The efficiency of conversion of ingested food (ECI) was calculated based on Waldbauer [24] as ECI = (DWG/FC) × 100 for each experimental unit.

Nutrient intake by *T. molitor* larvae was estimated from the self-selected consumption of the choice ingredients using the nutrient matrix calculation described by Morales-Ramos et al. [20,25]. The macro nutrient (lipid, protein and carbohydrate) content of the ingredients used in this study was obtained from data published in multiple sources [26–37]. The nutrient intake data was used to calculate the 3-way ratios of macro nutrient intake as described by Morales-Ramos et al. [25] and calculated as protein intake ratio = Pi/MNi, lipid intake ratio = Li/MNi, and carbohydrate intake ratio = Ci/MNi, where Pi, Li, and Ci are intakes of protein lipid and carbohydrate, respectively and MNi is the total intake of all three macronutrients and the sum of all three ratios is always = 1. The intake of other nutrients including neutral detergent fiber (NDF) and minerals including iron, magnesium, manganese, calcium, and zinc was also estimated.

Data consisting of live biomass gained, total food consumption, percent food assimilation, and efficiency of conversion of ingested food were compared among treatments using general linear mixed model (GLMM) and the Tukey–Kramer HSD (honestly significant difference) test for least square means of JMP software version 14.1 [38]. The effect of nutrient intake on food assimilation and efficiency of food conversion (ECI) was analyzed using multiple regression. The stepwise followed by backwards elimination methods were used to determine the optimal number of independent variables required in the model to explain food assimilation and ECI using the C_p statistic as the criterion to include or exclude variables [38–40].

3. Results

The means of consumption of each of the ingredients within each of the combination treatments are presented in Table 2. The relative consumption of each ingredient within combination treatments is illustrated as percentages in Figure 2. The ingredients that were consumed in higher proportion were dry potato in treatment 8 (41.01%); crude rice bran in treatments 3, 4, 5, 6, and 7 (40.47%, 34.87%, 30.37%, 32.27% and 33.18%, respectively); wheat bran in treatments 1 and 9 (30.49% and 37.1%, respectively); and corn dry distiller's grain with solubles (DDGS)in treatment 2 (34.63%). The least consumed ingredients were rice hulls in treatments 1 and 2 (0.31% and 0.13%, respectively); coffee chaff in treatment 8 (1.48%); peanut hulls in treatments 3, 6 and 7 (1.66%, 2.77%, and 3.19%, respectively); soybean meal in treatment 9 (2.06%); olive meal in treatment 4 (2.27%); and sunflower meal in treatment 5 (2.58%) (Figure 2). In general, highly consumed ingredients had a high carbohydrate content. The ingredients consumed in low percentages generally contained high amounts of fiber at the expense of other nutrients, such as rice hulls, coffee chaff and peanut hulls or have a combination of high fiber and high protein contents like meals of olive, soybean and sunflower.

Table 2. Dry-weight consumption (g) of food ingredients by *T. molitor* larvae in nine self-selection treatment combinations of eight choices during a three-week period.

Ingredient	Treatment								
	1	2	3	4	5	6	7	8	9
White Cabbage	4.49 ± 0.91			3.85 ± 0.35			4.16 ± 0.61		4.14 ± 0.45
Potato	5.09 ± 0.63				6.09 ± 0.55			10.33 ± 0.93	
Alfalfa pellets	1.26 ± 0.22	1.07 ± 0.24						1.8 ± 0.33	1.46 ± 0.24
Wheat Bran	8.95 ± 1.1		6.99 ± 0.98			5.81 ± 0.79			11.37 ± 0.81
Rice bran whole			11.38 ± 0.98	9.64 ± 0.73	8.55 ± 0.88	9.69 ± 0.51	9.78 ± 2.09		8.67 ± 1.7
Rice bran defatted		4.23 ± 1.19		4.21 ± 0.72	3.16 ± 1.03	4.51 ± 0.89	4.59 ± 1.84		
Spelt Screenings			2.1 ± 0.35						
Corn DDGS	5.06 ± 1.09	9.99 ± 1.23			2.68 ± 0.65	2.82 ± 0.64	4.35 ± 0.82		
Wheat DDGS				1.17 ± 0.26		1.02 ± 0.16			2.1 ± 0.75
Barley brewery spent grain		8.58 ± 0.59	4.5 ± 0.59	4.11 ± 0.52	3.09 ± 0.89	4.14 ± 0.4	3.2 ± 0.43	5.31 ± 0.43	
Canola meal	3.99 ± 0.9	4.27 ± 0.85		2.92 ± 0.31	2.71 ± 0.3		1.4 ± 0.46		0.75 ± 0.35
Soybean Meal								1.13 ± 0.14	0.62 ± 0.07
Sunflower meal					0.73 ± 0.16			3.54 ± 1.03	
Olive meal				0.63 ± 0.11					
Cotton seed meal			1.6 ± 0.35						
Kelp meal			0.47 ± 0.05						
Oat hulls			0.62 ± 0.29			1.21 ± 0.22	1.05 ± 0.19	1.65 ± 0.36	1.55 ± 0.39
Peanut hulls	0.42 ± 0.19	0.47 ± 0.12	0.47 ± 0.18	1.12 ± 0.21	1.16 ± 0.31	0.83 ± 0.18	0.94 ± 0.38	1.06 ± 0.54	
Rice hulls	0.09 ± 0.03	0.04 ± 0.02							
Coffee shaft		0.2 ± 0.09						0.37 ± 0.14	
Total Consumption	29.36 ± 2.13	28.84 ± 1.36	28.13 ± 1.36	27.06 ± 0.9	28.16 ± 1.35	30.03 ± 1.32	29.48 ± 1.35	25.18 ± 2.12	30.66 ± 1.53

Mean ± standard deviation.

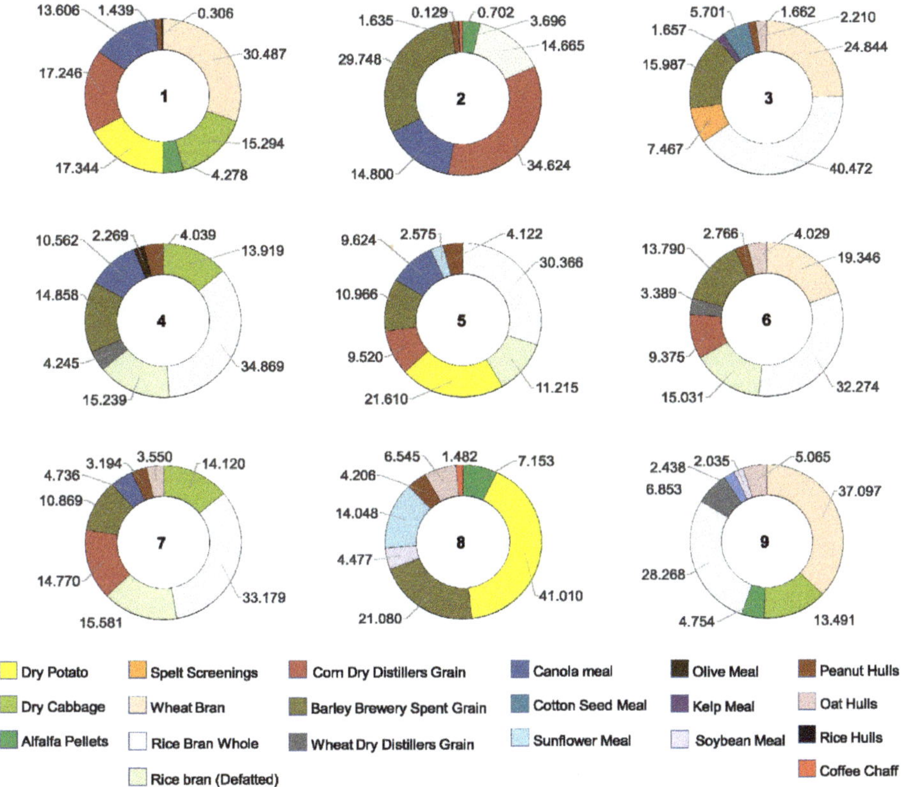

Figure 2. Proportional consumption of food ingredients by *T. molitor* larvae in nine treatments of different combinations of eight ingredients.

Despite the great diversity observed in the relative consumption of ingredients between treatments, the intake ratios of macro nutrients (lipid + protein + carbohydrate = 1) tended to converge close to a set of ranges between 0.03 to 0.16 of lipid, 0.21 to 0.25 of protein and 0.62 to 0.74 of carbohydrate (Table 3, Figure 3) with overall means of 0.1 ± 0.04, 0.24 ± 0.04, and 0.66 ± 0.06 for lipid, protein, and carbohydrate, respectively. The only exception was treatment 2 which showed significantly higher protein (0.36) ($F = 379.1$; df 8, 81; $p < 0.0001$) and lower carbohydrate (0.55) ($F = 460.8$; df 8, 81; $p < 0.0001$) intake ratio than all the other treatments (Table 3) outlying visibly in the graph of Figure 3. However, the rest of the treatments showed some significant differences among them in the macro nutrient intake ratios ($F = 304, 379.1$, and 460.8 for lipid, protein and carbohydrate, respectively; df 8, 81; $p < 0.0001$) that were less obvious in Figure 3 (Table 3).

Table 3. Macro nutrient intake ratios of *T. molitor* larvae in nine self-selection treatments with different combinations of eight food ingredients.

Treatment	Lipid	Protein	Carbohydrate
1	0.047 ± 0.003 [e]	0.23 ± 0.007 [c]	0.723 ± 0.008 [b]
2	0.092 ± 0.003 [d]	0.359 ± 0.006 [a]	0.549 ± 0.008 [g]
3	0.157 ± 0.006 [a]	0.206 ± 0.005 [d]	0.637 ± 0.003 [e]
4	0.132 ± 0.006 [b]	0.244 ± 0.005 [b]	0.624 ± 0.006 [f]
5	0.112 ± 0.005 [c]	0.22 ± 0.012 [c]	0.668 ± 0.01 [d]
6	0.134 ± 0.004 [b]	0.23 ± 0.004 [c]	0.636 ± 0.005 [e]
7	0.132 ± 0.017 [b]	0.23 ± 0.008 [c]	0.638 ± 0.014 [e]
8	0.029 ± 0.002 [f]	0.23 ± 0.008 [c]	0.741 ± 0.009 [a]
9	0.101 ± 0.011 [d]	0.205 ± 0.009 [d]	0.694 ± 0.009 [c]

Mean ± standard deviation. Means with the same letter are not significantly different after Tukey–Kramer HSD test at $\alpha = 0.05$.

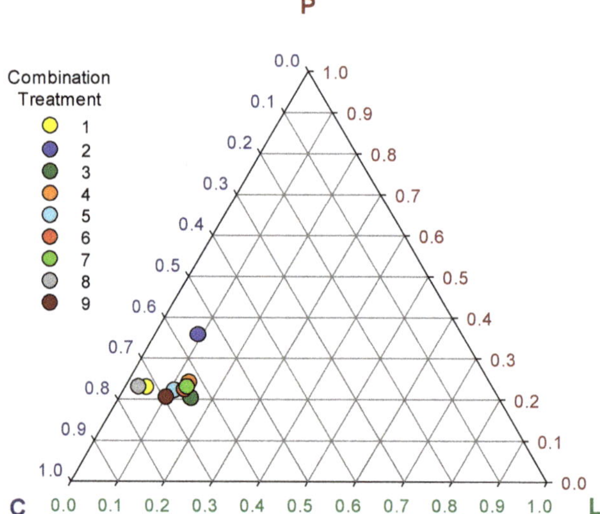

Figure 3. Ternary plot of self-selected macro-nutrient intake ratios (L = lipid, P = protein, C = carbohydrate) in nine combination treatments of eight ingredients.

These differences in the intake ratios of macro nutrients resulted in significant differences in group live biomass gain ($F = 10.15$; df = 8, 81; $p < 0.0001$), overall dry-weight food consumption ($F = 10.91$; df = 8, 81; $p < 0.0001$), food assimilation ($F = 29.13$; df = 8, 81; $p < 0.0001$), and ECI ($F = 28.41$; df = 8, 81; $p < 0.0001$) among choice treatments (Figure 4). The highest live biomass gain was observed in treatment 5 (7.3 ± 0.28 g), followed by treatments 1 (6.91 ± 0.41 g) and 7 (6.83 ± 0.74 g). The highest assimilation was observed in treatment 8 (55.25 ± 2.03%) followed by treatment 5 (50.86 ± 1.86%). The highest ECI was observed in treatment 5 (9.87 ± 0.45%) followed by treatment 8 (9.48 ± 0.64%). In general, the best performing treatments were 5, 8, and 1 (Figure 4). Treatment 2 was the worst performer among choice treatments, showing the lowest live biomass gain (5.45 ± 0.49 g), the lowest food assimilation (39.19 ± 2.11%), and the lowest ECI (7.18 ± 0.5%) (Figure 4). The low performance of larvae groups of treatment 2 may be associated with the significant deviations in macronutrient intake ratios observed in this treatment (Figure 3). The optimal macro-nutrient ratios for *T. molitor* may be closer to those observed in average for treatments 1, 5, and 8, which were 0.06 ± 0.03, 0.23 ± 0.01, and 0.71 ± 0.03 for lipid, protein, and carbohydrate, respectively.

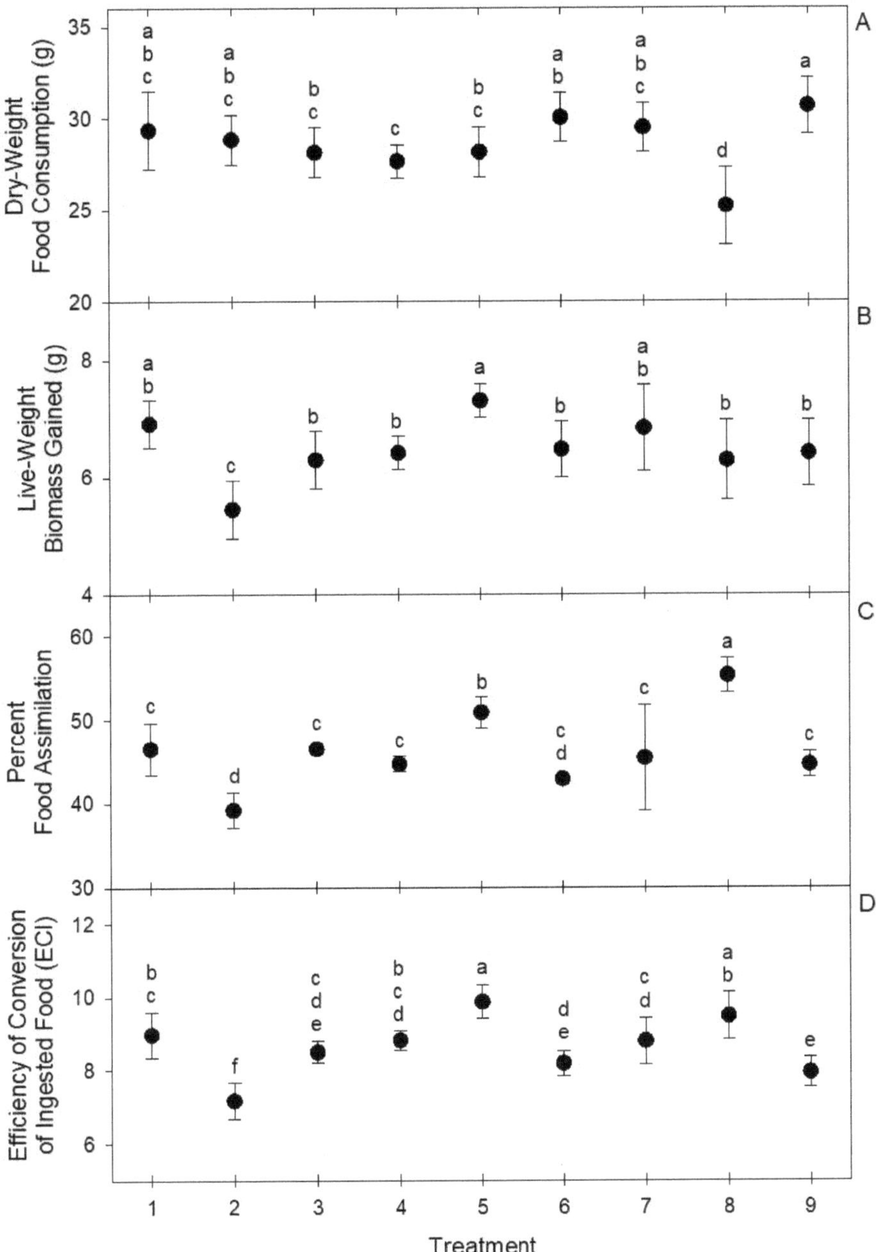

Figure 4. Circles represent means and brackets represent standard deviation of (**A**) dry-weight food consumption, (**B**) live biomass gain, (**C**) percent food assimilation, and (**D**) efficiency of conversion of ingested food (ECI) by groups of 100 *T. molitor* larvae in nine self-selection treatments of eight ingredients. Means with the same letter are not significantly different at α = 0.05 after Tukey–Kramer HSD test.

Live biomass gain was significantly impacted by efficiency of food conversion (ECI) ($R^2 = 0.53$; $F = 100.74$; df = 1, 88; $p < 0.0001$) and food assimilation ($R^2 = 0.13$; $F = 13.35$; df = 1, 88; $p = 0.0004$) in a positive way. Consumption of some ingredients have significant effects on biomass gain, food assimilation and ECI. For instance, consumption of potato had a significant positive effect on food assimilation ($\beta = 0.01$; $R^2 = 0.57$; $F = 116.64$; df = 1, 88; $p < 0.0001$), but consumption of corn DDGS had the opposite effect on food assimilation ($\beta = -0.007$; $R^2 = 0.21$; $F = 23.0$; df = 1, 88; $p < 0.0001$).

Ingredients that had a mean consumption percentage of at least 10% in any given choice treatment were considered relevant ingredients (RI). Relevant ingredients included potato, cabbage, wheat bran, crude rice bran, defatted rice bran, corn DDGS, spent brewery DG, canola meal, and sunflower meal. Multiple regression analysis indicated that the consumption of all the relevant ingredients had a significant positive effect on live biomass gain ($R^2 = 0.7$; $F = 20.75$; df = 9, 80; $p < 0.0001$). Only consumption of potato, cabbage, rice bran whole, and spent brewery DG had a significant positive effect on food assimilation (partial F Ratios = 49.47, 12.17, 6.62, and 8.55; df = 9, 80; $p < 0.0001$, = 0.0008, = 0.0119, and = 0.0045, respectively). Significant negative effects on food assimilation were observed with consumption of canola and sunflower meals (partial F Ratios = 6.39 and 4.49; df 9, 80; $p = 0.0135$ and 0.0371, respectively). The resulting optimized model for assimilation (after stepwise) agreed with the full model analysis including the 6 variables that showed significant effects on food assimilation ($R^2 = 0.75$; $p = 42.6$; df = 6, 83; $p < 0.0001$). In the full model (9 independent variables) the efficiency of food conversion (ECI) was only affected significantly by the consumption of potato, and this effect was positive (partial F Ratio = 13.31; df 9, 80; $p = 0.0005$). However, when this model was analyzed with the stepwise method, an optimized 3-variable model resulted that included potato, rice bran, and canola meal all affecting ECI significantly and positively ($R^2 = 0.64$; $p = 49.9$; df = 3, 88; $p < 0.0001$). Significant quadratic effects on live biomass gain were observed from consumption of potato ($\beta_1 = 0.146$, $\beta_2 = -0.023$; $R^2 = 0.18$; $F = 9.47$; df = 2, 87; $p = 0.0002$), corn DDGS ($\beta_1 = 0.077$, $\beta_2 = -034$; $R^2 = 0.38$; $F = 26.2$; df = 2, 87; $p < 0.0001$) and spent brewery DG ($\beta_1 = -0.105$, $\beta_2 = -0.024$; $R^2 = 0.27$; $F = 16.02$; df 2, 87; $p < 0.0001$). Biomass gain was maximized at an intermediate level of consumption of these three ingredients.

Intake ratios of some nutrients had a significant impact on food assimilation and efficiency of food conversion (ECI). The optimal multiple regression models obtained after stepwise and backwards elimination procedures consisted of only 2 dependent variables explaining food assimilation and 4 variables explaining ECI. Models are valid only within the ranges observed for these variables, presented in Table 4. Food assimilation was impacted significantly by carbohydrate and neural detergent fiber ($R^2 = 0.73$; $F = 117.93$; df = 2, 87; $p < 0.0001$) (Table 5). These two variables also impacted ECI in addition to the minerals Mg and Mn ($R^2 = 0.73$; $F = 57.48$; df = 4, 85; $p < 0.0001$) (Table 6).

Table 4. Summarized estimated nutrient intake means and ranges in 90 self-selection observations from 9 different 8-choice combination treatments (in mg/100 mg).

Nutrient	Mean ± SD	Minimum	Maximum
Lipid	8.13 ± 3.13	2.27	13.03
Protein	18.75 ± 2.78	15.58	26.45
Carbohydrate	51.94 ± 6.54	38.71	64.84
Fiber (ND)	28.06 ± 3.48	21.72	36.41
Ca	0.445 ± 0.246	0.11	0.906
Fe	0.014 ± 0.002	0.009	0.018
Mg	0.439 ± 0.113	0.157	0.568
Zn	0.004 ± 0.0007	0.002	0.005
Mn	0.007 ± 0.002	0.002	0.01

Table 5. Model from stepwise on percent assimilation.

Parameter	Estimate	Sum of Squares	F Ratio	$p > F$
Carbohydrate	$5.4 \times 10^{-4} \pm 4.3 \times 10^{-5}$	1121.48	158.76	<0.0001
Fiber	$-7.4 \times 10^{-4} \pm 8.1 \times 10^{-5}$	584.35	82.72	<0.0001

Model: $R^2 = 0.731$; F = 117.93; df 2, 87; $p < 0.0001$.

Table 6. Model from Stepwise and backwards elimination on ECI.

Parameter	Estimate	Sum of Squares	F Ratio	$p > F$
Carbohydrate	$1.2 \times 10^{-4} \pm 1.0 \times 10^{-5}$	27.22	120.17	<0.0001
Fiber	$-1.6 \times 10^{-4} \pm 1.5 \times 10^{-5}$	24.88	109.83	<0.0001
Mg	0.017 ± 0.003	6.14	27.1	<0.0001
Mn	-0.667 ± 0.14	4.14	22.71	<0.0001

Model: $R^2 = 0.73$; F = 57.48; df 4, 85; $p < 0.0001$.

4. Discussion

It is apparent by the results presented in this study that *T. molitor* larvae tend to balance their intake of macro nutrients by selecting among a variety of ingredients when feeding. This agrees with previous studies confirming the ability of *T. molitor* to self-select for optimal macro-nutrient intake ratios [41–44]. The intake ratios of macro nutrients by *T. molitor* larvae converged within a narrow range of values among eight of the nine combination treatments of different food ingredients. Treatment 2 was the exception showing excess intake of protein and reduced intake of carbohydrate. Deviation of macro-nutrient intake ratios observed in treatment 2 coincided with a low performance of growth and food utilization of the larvae grown in this treatment. The reason for the deviations in macro-nutrient intake ratios observed in treatment 2 may have been the absence of an additional ingredient with low protein content besides defatted rice bran. There was an unusually high consumption of corn DDGS (34.66 ± 4.13%) and spent brewery DG (29.75 ± 1.46%) in this treatment resulting in a combined mean consumption of 64.41% of these two ingredients from the mean total food consumption in treatment 2. In the other three treatments where these two ingredients were present together (treatments 5, 6, and 7), their combined consumption did not exceed 26% of the total food consumed. Additionally, consumption of corn DDGS and spent brewery DG did not exceed 21.5% when presented alone within the food choices (treatments 1, 3, 4, and 8). The high consumption of these two distilled grain ingredients in treatment 2 is itself an anomaly and may have been driven by the need for lipid intake, which was extremely low (lower than 3.6%) in the rest of the ingredients presented in treatment 2: two defatted ingredients (canola meal and rice bran defatted), alfalfa pellets, the hulls of peanut and rice, and coffee chaff [26,27,33,34]. The lipid content of corn DDGS and spent brewery DG is reported to be higher than 8% [26,27,30,34,35,37].

The optimal macro nutrient ratios for *T. molitor* may be those observed in the best performing treatments (1, 5, and 8): 0.06 ± 0.03 (max 0.12 min 0.03), 0.23 ± 0.01 (max 0.25 min 0.2), and 0.71 ± 0.03 (max 0.75 min 0.65) for lipid, protein, and carbohydrate, respectively. Rho and Lee (2016) [45] determined that an equal ratio of protein and carbohydrate was the best for *T. molitor* based on adult fecundity and longevity. However, this study is not comparable with ours because both studies were done on different life stages and measured different life cycle parameters.

Ingredients that were considered relevant based on relative consumption percentage (over 10%) included potato, cabbage, wheat bran, crude rice bran, defatted rice bran, corn DDGS, spent brewery DG, canola meal, and sunflower meal. Multiple regression analyses of consumption of relevant ingredients versus live biomass gain showed significant positive effects. These results can be interpreted as evidence that such ingredients are suitable for inclusion in diets for *T. molitor*, especially when biomass production is one of the main priorities. However, consumption of relevant ingredients did not always have positive effects on food assimilation. For instance, canola and sunflower meals had significant negative effects on assimilation. Food assimilation is not necessarily critical for biomass production when the food provided has a low cost, as in this case where agricultural by-products are used. Analysis of nutrient intake ratios showed that intake of fiber negatively affects food assimilation. This may explain the negative effects of canola and sunflower meals on assimilation, since both meals have a relatively high fiber content. Food conversion efficiency (ECI) was impacted positively by the consumption of potato, rice bran and canola meal. Because both food assimilation and ECI significantly impacted biomass gain in a positive way, we may consider the ingredients that impact both parameters in a positive way as highly suitable for inclusion in insect diets. Potato, rice bran, cabbage, spent brewery DG, and canola meal seem to be highly suitable as ingredients in *T. molitor* diets, but defatted rice bran, corn DDGS, and sunflower meal are promising if provided in the correct proportions. Wheat bran, potato, and cabbage have been used and are currently used regularly in *T. molitor* diets for mass production [46]. The rest of the ingredients are not currently used in commercial production, but some studies have assessed their potential, such as on spent brewery DG [32].

Macro-nutrient intake ratios were an important factor affecting live biomass gain, food assimilation and ECI. Macro-nutrient ratios were optimal for *T. molitor* within ranges of 0.06 ± 0.03, 0.23 ± 0.01, and 0.71 ± 0.03 for lipid, protein, and carbohydrate, respectively. Nutrient intake analyses showed that the intake of carbohydrate significantly and positively impacted live biomass gain, food assimilation and ECI. The intake of protein did not impact these three parameters within the ranges observed in this study. It appears that protein intake was strongly regulated by self-selection in most treatments, with the only exception of treatment 2. Other studies have reported that high protein intake reduce development time and pupal size [42] and increased adult longevity and fecundity [43]. In this study the impact of high intake of protein on biomass productivity and food utilization was negative. High intake levels of fiber also had a negative impact on food assimilation and ECI. Li et al. (2015) [47] reported that the optimal intake levels of crude fiber for *T. molitor* is within a range of 5 to 10%. In this study we did not compare intakes of crude fiber, but the self-selected percentages of ND fiber were between 22.52 ± 0.62% in treatment 5 and 34.94 ± 0.94% in treatment 9.

5. Conclusions

The macro-nutrient intake ratios resulting from ingredient self-selection by *T. molitor* fell within narrow margins: Lipid intake was between 0.12 and 0.03, protein between 0.25 and 0.2, and carbohydrate between 0.75 and 0.65. Deviations from these ranges of macro nutrient intake ratios resulted in a diminished performance in larval growth and food utilization.

The relevant ingredients, based on their relative consumption by *T. molitor* larvae included potato, cabbage, wheat bran, crude rice bran, defatted rice bran, corn DDGS, spent brewery DG, canola meal, and sunflower meal. Consumption of relevant ingredients significantly affected live biomass production in a positive way in *T. molitor* larvae.

Both food assimilation and efficiency of conversion of ingested food were positively impacted by ingestion of carbohydrate and negatively impacted by ingestion of fiber. Ingredients that enhanced both of these parameters had relatively high carbohydrate and low fiber content such as potato. However, levels of carbohydrate and fiber should not depart from the self-selected ranges observed, because excessive or deficient intake of those nutrients can have a detrimental impact on growth and food utilization in *T. molitor* larvae.

Author Contributions: Conceptualization, J.A.M.-R., M.G.R. and H.C.K.; methodology, J.A.M.-R., M.G.R. and H.C.K.; software, J.A.M.-R.; validation, J.A.M.-R. and H.C.K.; formal analysis, J.A.M.-R.; investigation, J.A.M.-R. and M.G.R.; resources, J.A.M.-R., M.G.R., H.C.K. and V.E.; data curation, J.A.M.-R.; writing—J.A.M.-R. and M.G.R.; writing—review and editing, J.A.M.-R., M.G.R., H.C.K. and V.E.; visualization, J.A.M.-R.; supervision, J.A.M.-R.; project administration, M.G.R. and V.E.; funding acquisition, V.E. All authors have read and agreed to the published version of the manuscript.

Funding: This research was funded in part by USDD-DARPA, SBIR Grant No. SB172-002, D2-2236.

Acknowledgments: We thank Damian Tweedy for his assistance in mixing and forming ingredient pellets.

Conflicts of Interest: The authors declare no conflict of interest. The funders had no role in the design of the study; in the collection, analyses, or interpretation of data; in the writing of the manuscript, or in the decision to publish the results.

References

1. Ng, W.-K.; Liew, F.-L.; Ang, L.-P.; Wong, K.-W. Potential of mealworm (*Tenebrio molitor*) as an alternative protein source in practical diets for African catfish, *Clarias gariepinus*. *Aquac. Res.* **2001**, *32*, 272–280. [CrossRef]
2. Barroso, F.G.; de Haro, C.; Sánchez-Muros, M.J.; Venegas, E.; Martínez-Sánchez, A.; Pérez-Bañon, C. The potential of various insect species for use as food for fish. *Aquaculture* **2014**, *422–423*, 193–201. [CrossRef]
3. Gasco, L.; Henry, M.; Piccolo, G.; Marono, S.; Gai, F.; Renna, M.; Lussiana, C.; Antonopoulou, E.; Mola, P.; Chatzifotis, S. *Tenebrio molitor* meal in diets for European sea bass (*Dicentrarchus labrax* L.) juveniles: Growth performance, whole body composition and in vivo apparent digestibility. *Anim. Feed Sci. Technol.* **2016**, *220*, 34–45. [CrossRef]
4. Gasco, L.; Gai, F.; Maricchiolo, G.; Genovese, L.; Ragonese, S.; Bottari, T.; Caruso, G. Fishmeal alternative protein sources for aquaculture. In *Feeds for the Aquaculture Sector*; Gasco, L., Gai, F., Maricchiolo, G., Genovese, L., Ragonese, S., Bottari, T., Caruso, G., Eds.; Springer Briefs in Molecular Science; Springer: Cham, Switzerland, 2018; pp. 1–28.
5. Lock, E.J.; Biancarosa, I.; Gasco, L. Insects as raw materials in compound feed for aquaculture. In *Edible Insects in Sustainable Food Systems*; Halloran, A., Flore, R., Vantomme, P., Roos, N., Eds.; Springer: Cham, Switzerland, 2018; pp. 263–276.
6. Ferrer, L.P.; Kallas, Z.; de Amores Gea, D. The use of insect meal as a sustainable feeding alternative in aquaculture: Current situation, Spanish consumers' perceptions and willingness to pay. *J. Clean. Prod.* **2019**, *229*, 10–21. [CrossRef]
7. Ramos-Elorduy, J.; Avila Gonzalez, E.; Rocha Hernandez, A.; Pino, J.M. Use of *Tenebrio molitor* (Coleoptera, Tenebrionidae) to recycle organic wastes and as feed for broiler chickens. *J. Econ. Entomol.* **2002**, *95*, 214–220. [CrossRef] [PubMed]
8. van Huis, A.; Van Itterbeeck, J.; Klunder, H.; Mertens, E.; Halloran, A.; Muir, G.; Vantomme, P. *Edible Insects, Future Prospects for Food and Feed Security*; FAO, ONU: Rome, Italy, 2013.
9. Makkar, H.P.S.; Tran, G.; Heuzé, V.; Ankers, P. State-of-the-art on use of insects as animal feed. *Anim. Feed Sci. Technol.* **2014**, *197*, 1–33. [CrossRef]
10. De Marco, M.; Martínez, S.; Hernandez, F.; Madrid, J.; Gai, F.; Rotolo, L.; Belforti, M.; Bergero, D.; Katz, H.; Dabbou, S.; et al. Nutritional value of two insect larval meals (*Tenebrio molitor* and *Hermetia illucens*) for broiler chickens: Apparent nutrient digestibility, apparent ileal amino acid digestibility and apparent metabolizable energy. *Anim. Feed Sci. Technol.* **2015**, *209*, 211–218. [CrossRef]
11. Sánchez-Muros, M.J.; Barroso, F.G.; Manzano-Agugliaro, F. Insect meal as renewable source of food for animal feeding: A review. *J. Clean. Prod.* **2014**, *65*, 16–27. [CrossRef]
12. Sánchez-Muros, M.J.; Barroso, F.G.; de Haro, C. Brief summary of insect usage as an industrial animal feed/feed ingredient. In *Insects as Sustainable Food Ingredients–Production, Processing and Food Applications*; Dossey, A.T., Morales-Ramos, J.A., Rojas, M.G., Eds.; Academic Press: San Diego, CA, USA, 2016; pp. 273–309.
13. Benzertiha, A.; Kierończyk, B.; Rawski, M.; Józefiak, A.; Kozłowski, K.; Jankowski, J.; Józefiak, D. *Tenebrio molitor* and *Zophobas morio* full-fat meals in broiler chicken diets: Effects on nutrients digestibility, digestive enzyme activities, and cecal microbiome. *Animals* **2019**, *9*, 1128. [CrossRef]
14. Gasco, L.; Biasato, I.; Dabbou, S.; Schiavone, A.; Gai, F. Animals fed insect-based diets: State-of-the-art on digestibility, performance and product quality. *Animals* **2019**, *9*, 170. [CrossRef]
15. Dossey, A.T.; Tatum, J.T.; McGill, W.L. Modern insect-based food industry: Current status, insect processing technology, and recommendations moving forward. In *Insects as Sustainable Food Ingredients–Production, Processing and Food Applications*; Dossey, A.T., Morales-Ramos, J.A., Rojas, M.G., Eds.; Academic Press: San Diego, CA, USA, 2016; pp. 113–152.

16. van Broekhoven, S.; Oonincx, D.G.A.B.; van Huis, A.; van Loon, J.J.A. Growth performance and feed conversion efficiency of three edible mealworm species (Coleoptera: Tenebrionidae) on diets composed of organic by-products. *J. Ins. Physiol.* **2015**, *73*, 1–10. [CrossRef] [PubMed]

17. Kim, S.Y.; Kim, H.G.; Lee, K.Y.; Yoon, H.J.; Kim, N.J. Effects of brewer's spent grain (BSG) on larval growth of mealworms, *Tenebrio molitor* (Coleoptera: Tenebrionidae). *Int. J. Ind. Entomol.* **2016**, *32*, 41–48. [CrossRef]

18. Zhang, X.; Tang, H.; Chen, G.; Qiao, L.; Li, J.; Liu, B.; Liu, Z.; Li, M.; Liu, X. Growth performance and nutritional profile of mealworms reared on corn stover, soybean meal, and distillers' grains. *Eur. Food Res. Technol.* **2019**, *245*, 2631–2640. [CrossRef]

19. Li, T.-H.; Che, P.-F.; Zhang, C.-R.; Zhang, B.; Ali, A.; Zang, L.-S. Recycling of spent mushroom substrate: Utilization as feed material for the larvae of the yellow mealworm *Tenebrio molitor* (Coleoptera: Tenebrionidae). *PLoS ONE* **2020**, *15*, e0237259. [CrossRef] [PubMed]

20. Morales-Ramos, J.A.; Rojas, M.G.; Dossey, A.T.; Berhow, M. Self-selection of food ingredients and agricultural by-products by the house cricket, *Acheta domesticus* (Orthoptera: Gryllidae): A Holistic approach to develop optimized diets. *PLoS ONE* **2020**, *15*, e0227400. [CrossRef] [PubMed]

21. Waldbauer, G.P.; Friedman, S. Self-selection of optimal diets by insects. *Annu Rev. Entomol.* **1991**, *36*, 43–63. [CrossRef]

22. Morales-Ramos, J.A.; Kelstrup, H.C.; Rojas, M.G.; Emery, V. Body mass increase induced by eight years of artificial selection in the yellow mealworm (Coleoptera: Tenebrionidae) and life history trade-offs. *J. Ins. Sci.* **2019**, *19*, 1–9. [CrossRef]

23. Morales-Ramos, J.A.; Kay, S.; Rojas, M.G.; Shapiro-Ilan, D.I.; Tedders, W.L. Morphometric analysis of instar variation in *Tenebrio molitor* (Coleoptera: Tenebrionidae). *Ann. Entomol. Soc. Am.* **2015**, *108*, 146–158. [CrossRef]

24. Waldbauer, G.P. The consumption and utilization of food by insects. *Adv. Ins. Physiol.* **1968**, *5*, 229–288.

25. Morales-Ramos, J.A.; Rojas, M.G.; Coudron, T.A. Artificial diet development for entomophagous arthropods. In *Mass Production of Beneficial Organisms: Invertebrates and Entomopathogens*; Morales-Ramos, J.A., Rojas, M.G., Shapiro-Ilan, D.I., Eds.; Academic Press: San Diego, CA, USA, 2014; pp. 203–240.

26. National Research Council (NRC). *United States-Canadian Tables of Feed Composition: Nutritional Data for United States and Canadian Feeds*, 3rd ed.; National Academy Press: Washington, DC, USA, 1982. Available online: http://nap.edu/1713 (accessed on 29 August 2019).

27. National Research Council (NRC). *Nutrient Requirements of Poultry*, 9th ed.; National Academy Press: Washington, DC, USA, 1994. Available online: http://nap.edu/2114 (accessed on 30 August 2019).

28. Wu, Y.V.; Stringfellow, A.C. Corn distillers' dried grains with solubles and corn distillers' dried grains: Dry fraction and composition. *J. Food Sci.* **1982**, *47*, 1155–1157. [CrossRef]

29. Batajoo, K.K.; Shaver, K.D. In Situ dry matter, crude protein, and starch degradabilities of selected grains and by-products. *Anim. Feed Sci Technol.* **1998**, *71*, 165–176. [CrossRef]

30. Spiehs, M.J.; Whitney, M.H.; Shurson, G.C. Nutrient database for distiller's dried grains with solubles produced from new ethanol plants in Minnesota and South Dakota. *J. Anim. Sci.* **2002**, *80*, 2639–2645. [PubMed]

31. Huzá, S.; Várhegyi, J.; Lehel, L.; Rózsa, L.; Kádár, M. Mineral content of grains, seeds and industrial by products. *Állattenyésztés Takarm.* **2003**, *52*, 277–283.

32. Kim, Y.; Moiser, N.S.; Hendrickson, R.; Ezeji, T.; Blaschek, H.; Dien, B.; Cotta, M.; Dale, B.; Ladisch, M.R. Composition of corn dry-grain ethanol by-products: DDGS, wet cake, and thin stillage. *Bioresour. Technol.* **2008**, *99*, 5165–5176. [CrossRef] [PubMed]

33. Kahlon, T.S. Rice bran: Production, composition, functionality and food applications, physiological benefits. In *Fiber Ingredients: Food Applications and Health Benefits*; Cho, S.S., Samuel, P., Eds.; CRC Press Taylor & Francis Group: Boca Raton, FL, USA, 2009; pp. 305–321.

34. Newkirk, R. *Canola Meal Feed Industry Guide*, 4th ed.; Canadian International Grains Institute, Canola Council of Canada: Winnipeg, MB, Canada, 2009.

35. U. S. Grains Council. *A guide to Distiller's Dried Grains with Solubles (DDGS)*, 3rd ed.; U. S. Grains Council: Washington, DC, USA, 2012.

36. US Department of Agriculture (USDA). *USDA National Nutrient Database for Standard Reference*; US Department of Agriculture: Beltsville, MD, USA, 2015. Available online: http://www.ars.usda.gov/ba/bhnrc/ndl (accessed on 13 October 2015).

37. Amorim, M.; Pereira, J.O.; Gomes, D.; Pereira, C.D.; Pinheiro, H.; Pintado, M. Nutritional ingredients from spent brewer's yeast obtained by hydrolysis and selective membrane filtration integrated in a pilot process. *J. Food Eng.* **2016**, *185*, 42–47. [CrossRef]

38. SAS Institute. *JMP Statistical Discovery from SAS, Version 14: Fitting Linear Models*; SAS Institute: Cary, NC, USA, 2018.

39. Myers, R.H. *Classical and Modern Regression with Applications*; Duxbury Press: Boston, MA, USA, 1986.

40. Freund, R.; Littell, R.; Creighton, L. *Regression Using JMP®*; SAS Institute Inc.: Cary, NC, USA, 2003.

41. Morales-Ramos, J.A.; Rojas, M.G.; Shapiro-Ilan, D.I.; Tedders, W.L. Self-selection of two diet components by *Tenebrio molitor* (Coleoptera: Tenebrionidae) larvae and its impact on fitness. *Environ. Entomol.* **2011**, *40*, 1285–1294. [CrossRef]

42. Morales-Ramos, J.A.; Rojas, M.G.; Shapiro Ilan, D.I.; Tedders, W.L. Use of nutrient self-selection as a diet refining tool in *Tenebrio molitor* (Coleoptera: Tenebrionidae). *J. Entomol. Sci.* **2013**, *48*, 206–221. [CrossRef]

43. Rho, M.S.; Lee, P. Geometric analysis of nutrient balancing in the mealworm beetle, *Tenebrio molitor* L. (Coleoptera: Tenebrionidae). *J. Ins. Physiol.* **2014**, *71*, 37–45. [CrossRef]

44. Rho, M.S.; Lee, P. Nutrient-specific food selection buffers the effect of nutritional imbalance in the mealworm beetle, Tenebrio molitor (Coleoptera: Tenebrionidae). *Eur. J. Entomol.* **2015**, *112*, 251–258. [CrossRef]

45. Rho, M.S.; Lee, P. Balanced intake of protein and carbohydrate maximizes lifetime reproductive success in the mealworm beetle, *Tenebrio molitor* (Coleoptera: Tenebrionidae). *J. Ins. Physiol.* **2016**, *91–92*, 93–99. [CrossRef]

46. Cortes Ortiz, J.A.; Ruiz, A.T.; Morales-Ramos, J.A.; Thomas, M.; Rojas, M.G.; Tomberlin, J.K.; Yi, L.; Han, R.; Giroud, L.; Jullien, R.L. Insect mass production technologies. In *Insects as Sustainable Food Ingredients–Production, Processing and Food Applications*; Dossey, A.T., Morales-Ramos, J.A., Rojas, M.G., Eds.; Academic Press: San Diego, CA, USA, 2016; pp. 154–201.

47. Li, L.; Stasiak, M.; Li, L.; Xie, B.; Fu, Y.; Gidzinski, D.; Dixon, M.; Liu, H. Rearing *Tenebrio molitor* in BLSS: Dietary fiber affects larval growth, development, and respiration characteristics. *Acta Astronaut.* **2015**, *118*, 130–136. [CrossRef]

Publisher's Note: MDPI stays neutral with regard to jurisdictional claims in published maps and institutional affiliations.

Article

Enterobacter sp. AA26 as a Protein Source in the Larval Diet of *Drosophila suzukii*

Katerina Nikolouli [1,2,*,†], Fabiana Sassù [1,2,3,†], Spyridon Ntougias [4], Christian Stauffer [2], Carlos Cáceres [1] and Kostas Bourtzis [1]

[1] Insect Pest Control Laboratory, Joint FAO/IAEA Centre of Nuclear Techniques in Food and Agriculture, Department of Nuclear Sciences and Applications, IAEA Laboratories, 2444 Seibersdorf, Austria; fabiana.sassu@gmail.com (F.S.); c.e.caceres-barrios@iaea.org (C.C.); k.bourtzis@iaea.org (K.B.)
[2] Department of Forest and Soil Sciences, Boku, University of Natural Resources and Life Sciences, 1190 Vienna, Austria; christian.stauffer@boku.ac.at
[3] Roklinka 224, Dolní Jirčany, 252 44 Psáry, Czech Republic
[4] Laboratory of Wastewater Management and Treatment Technologies, Department of Environmental Engineering, Democritus University of Thrace, Vas. Sofias 12, 67100 Xanthi, Greece; sntougia@env.duth.gr
* Correspondence: K.Nikolouli@iaea.org
† Equal Contribution.

Simple Summary: *Drosophila suzukii* has caused considerable damages to a variety of soft fruit crops. The sterile insect technique (SIT) is one of the most promising candidates that has recently attracted significant research efforts. A SIT package requires highly productive and cost-efficient mass rearing of the insects to produce a sufficient amount of males that will be sterilized and consequently released in the target area. Operational costs of mass-rearing facilities can be remarkably high, mainly due to the larval diet used for rearing. Gut symbiotic bacteria have been shown to enhance the productivity and development of fruit flies when used as supplements or protein source of their larval diet. In this study, we evaluated whether *Enterobacter* sp. AA26 could replace inactive brewer's yeast as a protein source in *D. suzukii* larval diet and effects on the biological quality of the flies are discussed.

Abstract: The Spotted-Wing Drosophila fly, *Drosophila suzukii*, is an invasive pest species infesting major agricultural soft fruits. *Drosophila suzukii* management is currently based on insecticide applications that bear major concerns regarding their efficiency, safety and environmental sustainability. The sterile insect technique (SIT) is an efficient and friendly to the environment pest control method that has been suggested for the *D. suzukii* population control. Successful SIT applications require mass-rearing of the strain to produce competitive and of high biological quality males that will be sterilized and consequently released in the wild. Recent studies have suggested that insect gut symbionts can be used as a protein source for *Ceratitis capitata* larval diet and replace the expensive brewer's yeast. In this study, we exploited *Enterobacter* sp. AA26 as partial and full replacement of inactive brewer's yeast in the *D. suzukii* larval diet and assessed several fitness parameters. *Enterobacter* sp. AA26 dry biomass proved to be an inadequate nutritional source in the absence of brewer's yeast and resulted in significant decrease in pupal weight, survival under food and water starvation, fecundity, and adult recovery.

Keywords: spotted-wing drosophila; symbiotic bacteria; gut microbiota; pest-management; mass-rearing; insect fitness

Citation: Nikolouli, K.; Sassù, F.; Ntougias, S.; Stauffer, C.; Cáceres, C.; Bourtzis, K. *Enterobacter* sp. AA26 as a Protein Source in the Larval Diet of *Drosophila suzukii*. *Insects* **2021**, *12*, 923. https://doi.org/10.3390/insects12100923

Academic Editors: Man P. Huynh, Kent S. Shelby and Thomas A. Coudron

Received: 30 August 2021
Accepted: 2 October 2021
Published: 9 October 2021

Publisher's Note: MDPI stays neutral with regard to jurisdictional claims in published maps and institutional affiliations.

1. Introduction

The gut of insects is the receptacle of a rich diversity of symbiotic bacteria, which can influence nutrient assimilation, host physiology, biology and ecology, including sexual and social behavior, and fitness [1–5]. The study of these microorganisms has brought to light mechanisms of symbiotic interactions that have proved to be a source of knowledge to manipulate insect behavior and develop microbe-based pest control techniques i.e., repellents, attract-and-kill, and mass trapping [6,7]. Recently gut microbiota have been exploited to increase the efficacy of the sterile insect technique (SIT) [8,9]. The SIT is an environment-friendly approach for population management with appreciable economic and social benefits that has been effectively applied worldwide on different insect pests [10,11]. SIT highly relies on efficient mass-rearing methods allowing the continuous production of large numbers of high-quality insects [12]. Insects (preferentially males) are then sterilized by irradiation and released with an overflow ratio into a target area, where they are expected to compete with wild males and mate with wild females [12]. Successful matings between sterile males and wild females will not produce viable offspring that eventually cause the intended population to decline in the next generations [10].

One of the most significant parts of a SIT programme is the larval diet used during insect mass-rearing. The artificial diet is one of the most critical and expensive components for the production of high-quality sterile insects and its development is guided by the availability of ingredients, nutritional value, ease of storage, labour and cost [13,14]. The diet used for mass-reared insects should supply proper and sufficient nutritional factors allowing for an efficient maturation into the adult stage and high adult fitness [13,15]. In addition, the components of the mass-rearing diet should be cost-effective without any side effects on insect fitness and overall colony productivity [13,16]. In the long run, diet expenses and insect performance should be well-balanced to achieve highly competitive insects at a reduced cost.

Several studies have focused on exploiting insect gut symbiotic bacteria as probiotics or alternative protein sources to enhance the SIT of fruit flies [17–28]. In the case of *Ceratitis capitata* (Wiedemann) (Diptera: Tephritidae), enrichment of the larval diet with live bacteria of the *Enterobacteriacae* family shortened the immature development stages, extended survival and improved the male mating competitiveness [9,17,21–24]. Increased fecundity was also observed when *Bactrocera oleae* (Rossi) (Diptera: Tephritidae) was fed with *Enterobacteriacae* microbiota [27,29]. Similarly, bacteria such as *Candidatus* Erwinia dacicola [30] and *Pseudomonas putida*, which are common in the wild population of *Bactrocera oleae* (Rossi) (Diptera: Tephritidae), were found beneficial for larval development and female fecundity, respectively [27,29].

In all the above studies the beneficial probiotic effect was shown by inoculating the diet with live bacteria while the use of inactive bacteria is limited [18,28]. Kyritsis and colleagues [28] assessed whether the use of inactive *Enterobacter* sp. AA26 biomass could be used in the larval diet of *C. capitata* as an alternative protein source to replace the costly brewer's yeast. Incorporating dead *Enterobacter* sp. AA26 benefited substantially the biological quality and productivity of reared *C. capitata* and paved the way to explore the use of inactive bacteria as the main protein source in the larval diet.

Drosophila suzukii (Diptera: Drosophilidae) is one of the most damaging insect pests of soft skinned fruits in North America and Europe in the last decade, but also a newly introduced detrimental insect in South America and Africa [31–34]. The also-called spotted wing Drosophila (SWD) fly can damage a wide range of economically important soft-skin fruits that are easily perforated by the females sclerotized ovipositor to lay their eggs [35,36]. Eggs hatch and larvae feed on the fruit pulp which causes the total rot of the fruit and an economic impact that accounts for millions of revenue losses each year [37]. *Drosophila suzukii* infestations are primarily controlled by the application of various insecticides, but the rising concerns on their poor effectiveness and possible impacts on the health of farmers and consumers have initiated the development of eco-friendly pest management tactics such as netting and tunneling, attract-and-kill baits deployment, alternative oviposition sites, and parasitoid releases [38,39]. Researchers have recently begun to evaluate SIT as a strategy to control *D. suzukii* populations in confined areas [40–42]. Currently, species-specific protocols of irradiation, mass-rearing, packaging, and quality control are fully or partially available for the implementation of this technique to control *D. suzukii* populations [42–44]. Nevertheless, the use of probiotics in *D. suzukii*'s larval or adult diet can help to further improve some of these protocols i.e., increase the mass-rearing productivity and/or boost the quality of the released sterile flies.

Previous studies have characterized bacterial species frequently occurring within the gut of *D. suzukii*, in some cases as an attempt to improve the effectiveness of bait traps used for its population management [45,46], and in others to understand its peculiar specialization in ripe and ripening fruits compared to other drosophilid species [47–50]. As a result, bacterial families such as *Enterobacteriaceae*, *Acetobacteraceae*, *Lactobacillaceae*, and *Enterococcaceae*, which are generally observed in association with different wild or laboratory-reared species of drosophilid [51], were also found in *D. suzukii* [52,53]. Several studies have identified *Enterobacteriaceae* to be abundant in *Drosophila melanogaster* [51,54,55] and *D. suzukii* [47,49,50,52,53,56]. Martinez-Sañudo et al. observed that *Enterobacteriaceae* in *D. suzukii* were more abundant and diverse in newly colonized areas compared to flies adapted in the new habitat [53]. Interestingly, *Enterobacter* sp. has been identified in few *D. suzukii* studies with varying diversity and abundance [52,53,57].

Distinct microbial populations have been correlated with positive impacts on fly longevity and development [49,58]. Bing et al. showed that *D. suzukii* utilizes microbes as a source of protein when reared on fruit-based diets [49]. During undernutrition *D. melanogaster* can use a wide range of microbes which serve as a protein source [58]. However, it is not yet clear how all fly-associated microbes contribute to nutrition.

A comparative study between *Enterobacter* sp. AA26 and Torula yeast (a yeast type that is different than brewer's yeast) showed that *Enterobacter* sp. AA26 was equivalent to yeast in terms of nutritional value [59]. The biomass of the strain AA26 provided all the essential and non-essential amino acids and vitamins required for the development of *C. capitata* larvae. Considering the abundance of *Enterobacteriaceae* species in fruit flies, including *D. suzukii*, the beneficial effects of *Enterobacter* sp. AA26 on the mass-reared *C. capitata*, the high protein content of *Enterobacter* sp. AA26 and the role of microbes as protein sources for *Drosophila* flies, we evaluated *Enterobacter* sp. AA26 as a potential protein source that could totally or partially replace inactive brewer's yeast in the *D. suzukii* larval diet. Assessing the potential beneficial effect of the same bacteria species in different insect species is of critical importance because if results are positive, they will justify economically the insect production in mass-rearing facilities. We assessed the effect of *Enterobacter* sp. AA26 on several fitness parameters, including pupal weight and recovery, adult emergence rate, sex ratio, flight ability, adult survival under stress and female fecundity. These fitness parameters are considered production and quality indicators of flies used for SIT programmes and ensure that insects of poor quality that can lead to a lack of effective control and higher programme costs are excluded [60].

2. Materials and Methods

2.1. Drosophila Suzukii Rearing Colony

All experiments were performed using the *D. suzukii* colony maintained at the Insect Pest Control Laboratory (IPCL) of the Joint FAO/IAEA Centre of Nuclear Techniques in Food and Agriculture, Seibersdorf, Austria. The colony was obtained from the Agricultural Entomology Unit of the Edmund Mach Foundation in San Michele All'Adige, Trento, Italy. Adult flies were maintained in metal-framed and mesh-covered cages of $45 \times 45 \times 45$ cm under controlled laboratory temperature, RH and light conditions (22 ± 5 °C, $65 \pm 5\%$ RH, and 14:10 h light:dark (L:D) photoperiod).

2.2. Enterobacter sp. AA26 Biomass Production and Larval Diet Preparation

Inactive *Enterobacter* sp. AA26 biomass was produced as described in Kyritsis et al. [28]. Briefly, a 1 L laboratory-scale bioreactor was used to grow *Enterobacter* sp. AA26 under aseptic conditions. The bioreactor was fed with Luria-Bertani (LB) broth and operated under the fill and draw mode. Adequate aeration of the bioreactor was achieved through an air pump and continuous agitation. The bacterial biomass was collected by centrifugation at $4000 \times g$ for 10 min and stored at -80 °C until use.

The biomass was dried at 60 °C for 24 h and subsequently grinded in a Planetary Ball Mill PM100 for 3.5 min, until a fine powder was obtained. The rearing larval diet, consisting of 28% wheat bran, 7% inactive brewer's yeast (*Saccharomyces cerevisiae*), 13% sugar, 0.45% sodium benzoate, 0.45% nipagin, and 51% water, was used as a standard control diet. All solid components (apart from nipagin) were weighted and mixed into water to a total volume of 1 L. The solution was constantly stirred and brought to boil. It was left simmering for 10 min and then cooled slightly for an additional 10 min. Nipagin was diluted in 10 mL water, added to the media and mixed well. To evaluate the effect of yeast replacement with *Enterobacter* sp. AA26, we used 1) a full replacement diet (hereafter as "total") containing 7% *Enterobacter* sp. AA26 biomass instead of inactive brewer's yeast, and a partial replacement diet with 3.5% inactive brewer's yeast and 3.5% *Enterobacter* sp. AA26 (hereafter as "partial"). We decided to use the 1:1 *Enterobacter* sp. AA26:yeast ratio based on the similar study performed in medfly [28].

2.3. Drosophila Suzukii Egg Collection

Eggs used for the experiments were obtained following the wax-rearing procedure developed at the IPCL [44]. Eggs were collected from the mass-rearing colony of the IPCL and placed on moist filter paper to avoid desiccation. All eggs were collected within a period of 4 h to prevent sample variation. Twenty-four hours after the collection, eggs were transferred on a wet black net in a Petri dish (size: 70 × 15 mm (D × H)) containing about 150 g of diet. Four replicates of 500 eggs each were collected per treatment (i.e., $n = 2000$ eggs per treatment). These eggs were used to evaluate pupal weight and recovery, adult emergence rate, sex ratio, flight ability, adult survival under stress and female fecundity. The number of replicates used in each experiment are shown below in the respective M&M section.

2.4. Effect of Enterobacter sp. AA26 on Pupae and Adult Recovery

To assess the pupae and adult recovery, we measured number of pupae, adult emergence and sex ratio. Ten days after the egg collection, all pupae were removed from the diet and counted. All pupae that recovered from the same larval treatment (4 petri dishes per treatment) were placed on moist paper and left in boxes until adult emergence in a room with regulated temperature, humidity and light. Adult emergence was recorded daily using CO_2 anesthesia. The sex ratio was determined as proportion of males per total number of adults. Adult emergence and sex ratio data were also collected from the flight ability and survival under stress experiments and all data coming from the different assays were combined for statistical analysis.

2.5. Effect of Enterobacter sp. AA26 on Pupal Weight

To determine the pupal weight, dark-brown pupae 24 h before emergence were selected from each treatment. Until the day of the experiment pupae were maintained in a room with regulated temperature, humidity and light to avoid any bias in the weight data. Pupal weight was determined by weighing independently pools of 10 pupae. Each single pool represented one sample. For each treatment (partial, total, and control), we performed 12 replicates ($n = 120$ pupae per treatment). All data were combined for statistical comparisons.

2.6. Effect of Enterobacter sp. AA26 on Survival

The adult survival was tested under food and water deprivation. Two days before emergence, 100 pupae from each treatment were randomly collected and individually placed into a 96-well microtiter plate sealed with plastic film. Pupae were checked twice a day to record the time of emergence, the time of death, and the sex of each fly. The film on the top of each well was delicately perforated to allow air exchange. Plates were kept in the dark to reduce the mobility at standard laboratory conditions. One plate was set up for each treatment. Each single adult represented one sample.

2.7. Effect of Enterobacter sp. AA26 on Flight Ability

The flight ability test was performed following the standard Quality Control Procedures applied to evaluate the sterile insects used in SIT applications. Pupae from each treatment were placed within a ring of paper centered in the bottom of the Petri dish to allow newly emerged flies a place to rest. A black plexiglass tube was placed over the Petri dish and was lightly coated inside with unscented talcum powder to prevent the flies from walking out. Flies that emerged were periodically removed from the vicinity of the tubes to minimize fly-back or fall-back into the tubes. For each treatment four replicates were conducted each consisting of 30 pupae ($n = 120$ pupae per treatment).

2.8. Effect of Enterobacter sp. AA26 on Female Fecundity

Newly emerged adults were selected from each treatment, sexed and separated into "triplets" made of one female and two males to ensure female insemination. Triplets were then transferred into 200 cm^3 volume rectangular plexiglass cages and were provided with water and standard adult diet under the standard laboratory conditions. Twenty replicates were performed for each of the treatments and the control group. After 72 h, males were discarded, and females were individually placed into a Petri dish provided with raspberry-juice agar substrate as egg oviposition site. Oviposition was allowed for 48 h without any interruption, after which the females were discarded, and the number of laid eggs was counted. Each single female represented a replicate. Females that did not lay any eggs or laid fewer than 10 eggs were not included in the analysis.

2.9. Statistical Analyses

All statistical analyses were performed using R version 4.0.5 [61].

Pupal weight: Pupal weight data are continuous variables and therefore assume a normal distribution. A linear model was applied for their analysis.

Pupae and adults recovery: The number of pupae and adults recovered per treatment are count data and were analyzed with GLM with negative binomial distribution and a log link function. Negative binomial was applied due to overdispersion detected in the Poisson and the Quasi-Poisson GLM models [62,63]. Analysis of deviance was performed with an F-test [64].

Sex ratio (proportion males): Sex ratio proportional data assume a binomial distribution and were analyzed with a GLM-binomial family and a logit link function [65]. Analysis of deviance was performed with a Chi-squared test [64].

Adult survival: The survivorship curves were calculated using a Kaplan–Meier approach (survfit package) [66]. The package survival was used for modeling the survival data [67].

Flight ability: Rate of fliers are proportional data and assume a binomial distribution [65]. Data were analyzed with a GLM-binomial family and a logit link function. Analysis of deviance was performed with a Chi-squared test [64].

Fecundity: Fecundity data are count data and were analyzed with a generalized linear model (GLM) with negative binomial distribution and a log link function. Negative binomial was applied due to overdispersion detected in the Poisson model [62,63]. Analysis of deviance was performed with an F-test [64].

Residuals of the models were checked for normality and homogeneity of variance. Goodness-of-fit of the models was visually inspected with half-normal plots with simulation envelopes [68]. In addition, analysis of variance was performed for each model to check assess differences in the fit statistics [65]. Overdispersion of the generalized linear models was checked with the DHARMA package [69]. DHARMA tests if the simulated dispersion is equal to the observed dispersion and supports the visual inspection of the residuals. Lsmeans package were used for the pair-wise comparisons of the fitted model estimates [70]. In all cases, the mean $\pm$ standard error is reported. In all boxplots, both the mean and the median values are depicted. Significant differences between treatment groups are indicated in the boxplots with asterisks (*** $p \leq 0.001$, ** $p \leq 0.01$, * $p \leq 0.05$, ns: $p > 0.05$; confidence level used: 0.95, alpha = 0.05). Non-significant differences are not shown in the boxplots.

3. Results

3.1. Effect of Enterobacter sp. AA26 on Pupal Weight

Brewer's yeast replacement with *Enterobacter* sp. AA26 had a significant effect on the average weight of the pupae (linear model: F = 26.45, df = 2, 33, $p = 1.395 \times 10^{-7}$) (average pupal weight: total: 0.0146 g $\pm$ 0.0007; partial: 0.0162 g $\pm$ 0.0006; control: 0.0166 g $\pm$ 0.0008). Pairwise comparisons among the tested diets (total, partial, control) indicated that the pupal weight was significantly lower in the total replacement when compared either with the partial or the control treatments (t-value = 5.351, $p \leq 0.0001$; and t-value = 6.942, $p \leq 0.0001$, respectively), while, between the partial and the control treatments, the difference was not significant (t-value = 1.591, $p = 0.1212$) (Figure 1).

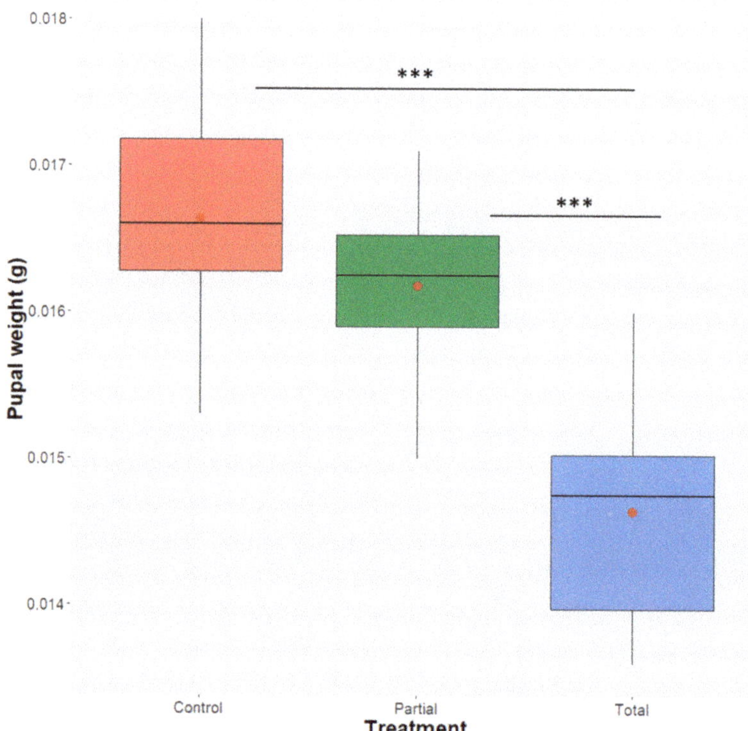

Figure 1. Effect of *Enterobacter* sp. AA26 on pupal weight. Data were analyzed with a linear model to define if the different treatments had a significant effect on the average pupal weight. Boxplots span the interquartile range and whiskers indicate the highest and lowest observations. The line and the dot inside each box represent the median and the mean, respectively. Significant differences between treatment groups are indicated with asterisks (*** $p \leq 0.001$, ** $p \leq 0.01$, * $p \leq 0.05$, ns: $p > 0.05$; confidence level used: 0.95, alpha = 0.05).

3.2. Effect of Enterobacter sp. AA26 on Developmental Parameters

3.2.1. Pupae Recovery

The mean number of pupae recovered per treatment grown in the total replacement diet was 261 $\pm$ 20.15 compared to 318 $\pm$ 31.2 and 352 $\pm$ 51.5 of the partial and control treatments, respectively (Figure 2a). Analysis of deviance indicated that treatment is not a significant predictor for pupal recovery (GLM negative binomial model: F = 2.0914, df = 2, 9, p = 0.123) based on the GLM-negative binomial model. A marginal significant difference was detected at the pairwise comparison of the total and partial treatments (z = 2.026, p = 0.0427).

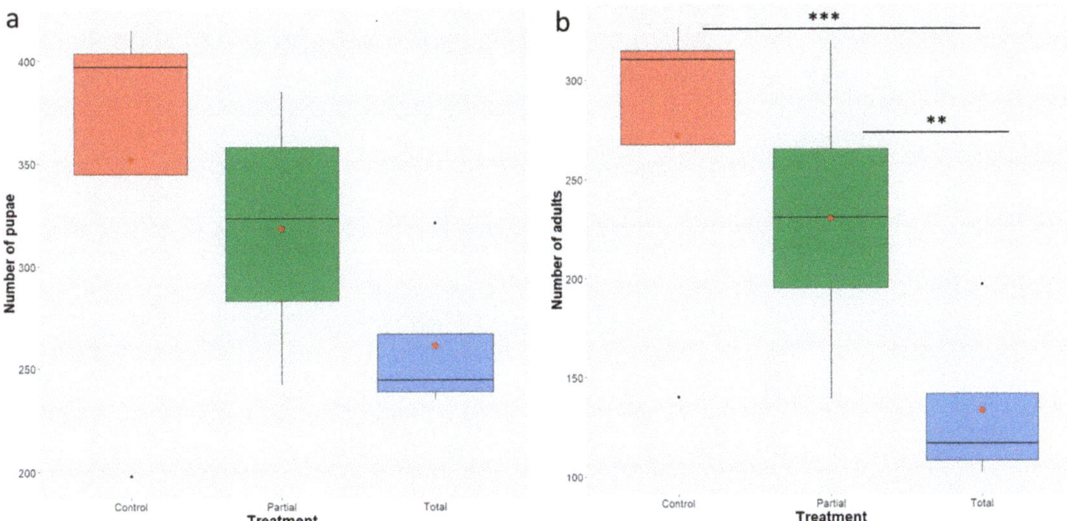

Figure 2. Effect of *Enterobacter* sp. AA26 on pupal and adult recovery. A GLM (negative binomial family) was used to analyze the impact of the various treatments on the recovery of pupae and adults; (**a**). Number of pupae recovered per treatment; (**b**). Number of adults recovered per treatment. Boxplots span the interquartile range and whiskers indicate the highest and lowest observations. The line and the dot inside each box represent the median and the mean, respectively. Significant differences between treatment groups are indicated with asterisks (*** $p \leq 0.001$, ** $p \leq 0.01$, * $p \leq 0.05$, ns: $p > 0.05$; confidence level used: 0.95, alpha = 0.05).

3.2.2. Adult Recovery and Sex Ratio

Analysis of deviance indicated that treatment is a significant predictor for the number of adults recovered (GLM negative binomial model: F = 5.9898, df = 2, 9, p = 0.002504). Total replacement of brewer's yeast with *Enterobacter* sp. AA26 significantly decreased the adult recovery compared to both the partial replacement and the control treatment (z = 2.610, p = 0.0090 and z = 3.416, p = 0.0006, respectively) (Figure 2b). We did not observe any significant adult recovery decrease between the partial replacement and the control treatment (z = 0.808, p = 0.4190) (Figure 2b).

The partial and total replacement of *Enterobacter* sp. AA26 did not affect the sex ratio of the emerged adults (GLM binomial model: analysis of deviance: χ^2 = 4.8549, df = 9, p = 0.218) (Figure 3).

3.3. *Effect of Enterobacter* sp. *AA26 on Adult Survival under Food and Water Starvation*

The presence of *Enterobacter* sp. AA26 had a negative impact on the *D. suzukii* adult survival under food and water starvation (♀-rank test: χ^2 = 19, df = 2, $p = 7 \times 10^{-5}$; ♂log-rank test: χ^2 = 40.5, df = 2, $p = 2 \times 10^{-9}$). Males developed in the total replacement diet had significantly shorter survival times compared to the ones developed in the partial and the control diets (log-rank test: χ^2 = 5.7, df = 1, p = 0.02; log-rank test: χ^2 = 15.8, df = 1, $p = 7 \times 10^{-5}$, respectively). In the case of females, the survival probability was significantly shorter when grown in the total replacement diet compared to the control one (log-rank test: χ^2 = 18.9, df = 1, $p = 1 \times 10^{-5}$), but no significant difference was detected between the partial and the control diets (log-rank test: χ^2 = 0.4, df = 1, p = 0.5) (Figure 4).

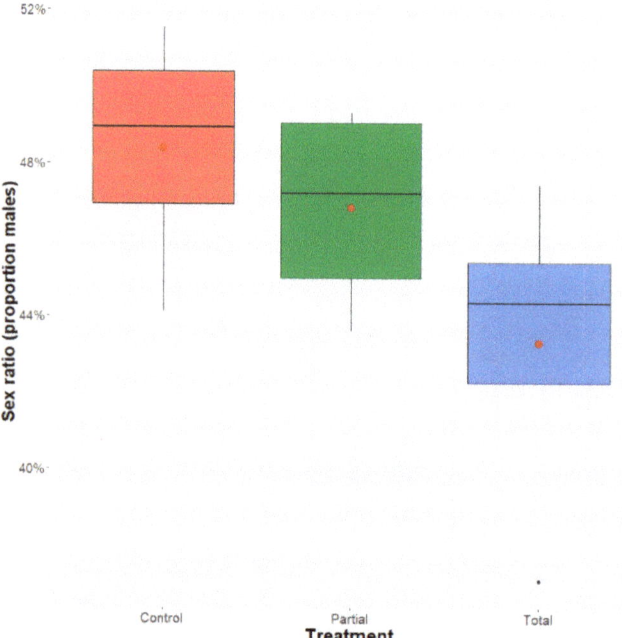

Figure 3. Sex ratio of males developed in the partial and total replacement of *Enterobacter* sp. AA26 diet and control diet. Sex ratio was determined as the percentage of males per total number of adults. A GLM (binomial family) analysis was performed to determine the effect of *Enterobacter* sp. AA26 diet replacement. Boxplots span the interquartile range and whiskers indicate the highest and lowest observations. The line and the dot inside each box represent the median and the mean, respectively. Significant differences between treatment groups are indicated with asterisks (*** $p \leq 0.001$, ** $p \leq 0.01$, * $p \leq 0.05$, ns: $p > 0.05$; confidence level used: 0.95, alpha = 0.05).

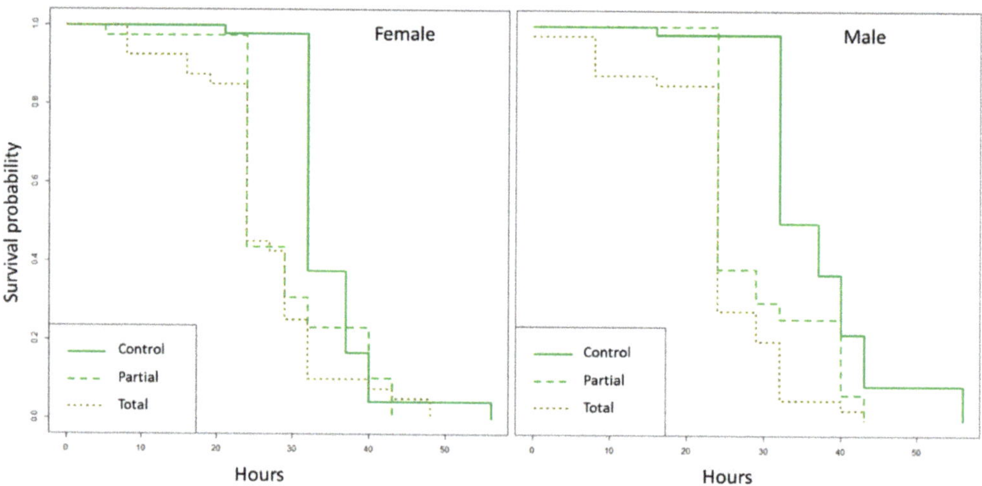

Figure 4. Effect of *Enterobacter* sp. AA26 on female (**left**) and male (**right**) survival under starvation. Significant differences were measured with a log-rank test. The x-axis represents time in hours.

3.4. Effect of Enterobacter sp. *AA26 on Flight Ability*

The addition of *Enterobacter* sp. AA26 in the *D. suzukii* larval diet did not affect the adult flight ability (GLM binomial model-analysis of deviance: $\chi^2 = 14.829$, df = 9, $p = 0.1942$). The rate of fliers was calculated based on the emerged pupae and data showed that there was no significant difference among the three treatments. The rate of fliers was 76.3% ± 1.28, 77.7% ± 0.80, and 85% ± 0.80 for the total, partial and control treatments, respectively (Figure 5).

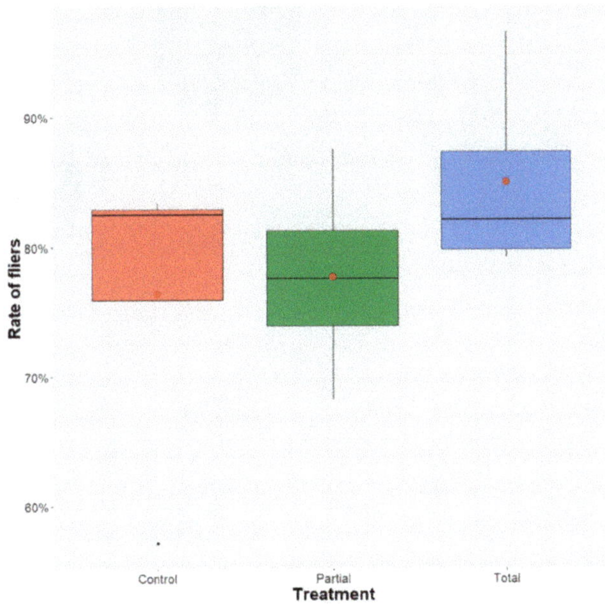

Figure 5. Flight ability of *D. suzukii* adults. The rate of fliers was calculated based on the emerged pupae. A GLM (binomial family) analysis was performed to determine the effect of *Enterobacter* sp. AA26 diet replacement. Boxplots span the interquartile range and whiskers indicate the highest and lowest observations. The line and the dot inside each box represent the median and the mean, respectively. Significant differences between treatment groups are indicated with asterisks (*** $p \leq 0.001$, ** $p \leq 0.01$, * $p \leq 0.05$, ns: $p > 0.05$; confidence level used: 0.95, alpha = 0.05).

3.5. Effect of Enterobacter sp. *AA26 on Fecundity*

Analysis of deviance indicated that treatment is a significant predictor for the egg production of *D. suzukii* females (GLM negative binomial model: F = 15.263, df = 2, 37, $p = 2.352 \times 10^{-7}$). The mean number of eggs per female grown in the total replacement diet (22.1 ± 3.88) was significantly lower compared to the partial (35.9 ± 6.76) and the control diets (44.7 ± 6.01) (total-partial: z = 3.299, $p = 0.0010$; total-control: z = 5.516, $p < 0.0001$). The model did not detect a significant difference between the partial and the control treatments (z = 1.848, $p = 0.0646$) (Figure 6).

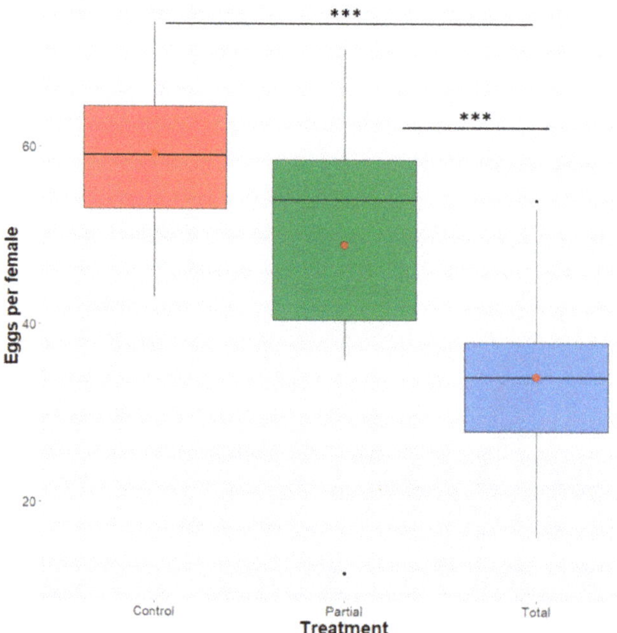

Figure 6. Effect of *Enterobacter* sp. AA26 on *D. suzukii* female fecundity. The number of eggs per female are shown in the y-axis for each of the treatment groups and control group. Analysis of fecundity were performed using a GLM (negative binomial family). Boxplots span the interquartile range and whiskers indicate the highest and lowest observations. The line and the dot inside each box represent the median and the mean, respectively. Significant differences between treatment groups are indicated with asterisks (*** $p \leq 0.001$, ** $p \leq 0.01$, * $p \leq 0.05$, ns: $p > 0.05$; confidence level used: 0.95, alpha = 0.05).

4. Discussion

In our study, results indicate that *Enterobacter* sp. AA26 dry biomass is not adequate to fully replace inactive brewer's yeast as a protein source in *D. suzukii* larval diet. In particular, complete replacement of brewer's yeast resulted in significant decrease in pupal weight, survival under food and water starvation, fecundity, and adult recovery. In addition, neither the partial nor the complete replacement of yeast with *Enterobacter* sp. AA26 had any significant impact on flight ability, sex ratio, and pupal recovery.

Core aspects of insect physiology are influenced or regulated by gut microbial communities by promoting digestive activities, boosting immune responses and restricting pathogen colonization [71]. Several studies have characterized the diverse microbial communities harboring the gut of natural *Drosophila* populations, with *Enterobacteriaceae* being one of the predominant families [51,72,73]. Studies on *D. suzukii* also confirmed the presence of *Enterobacteriaceae* in their digestive tract [52,53]. *Enterobacter* belongs to the *Enterobacteriaceae* family and is considered one of the most dominant genera of the gut of several insect species [74–77]. Due to their pivotal role in host physiology and biology, *Enterobacter* spp. could be exploited for the control of pest species. Recent studies have explored the possibility of using *Enterobacter* spp. as probiotic supplement in larval diets of mass-rearing insects for large scale operational SIT programs [22–24,28]. *Enterobacter* sp. AA26 was isolated from the gut of *C. capitata* males and females and assessed for its probiotic effects on several fitness parameters [23,24]. Results in *C. capitata* indicated that addition of *Enterobacter* sp. AA26 increased pupae and adult production and decreased rearing duration for several developmental stages. On the other hand, pupal weight, sex ratio, male mating competitiveness, flight ability or life span under food and water deprivation were not affected. *Enterobacter* sp. AA26 dry biomass was also tested as a potential protein source that could fully replace the brewer's yeast in the *C. capitata* larval diet [28]. *Enterobacter* sp. AA26 proved to be an adequate nutritional source for *C. capitata* larvae since the immature stages' mortality decreased, the pupae development was accelerated, the pupal weight increased and the adult survival under food and water deprivation was elongated.

Drosophila suzukii is a continuously expanding threat and a SIT-based approach has been proposed as a promising control option. Mass-rearing protocols and the larval diet are important aspects that will determine the production of insects of high biological quality that will be released in the target area. *Drosophila suzukii* diet is currently based on brewer's yeast as a protein source. Previous studies have pointed out the critical role of yeast for *D. suzukii* as a source of dietary protein that is absent from fruits [78–81]. Yeast has been proved to have an essential nutritional role in *D. suzukii* larval development, survivorship, eclosion rate, and adult body mass [82–84]. The mass-rearing protocol of *D. suzukii* is currently based on a wheat bran larval diet that includes 7% inactive brewer's yeast [44]. Information collected from SIT facilities, currently rearing *C. capitata*, indicate that yeast-related expenses can be very high and, in some cases, they can be as high as 12% of the whole production cost [28]. Therefore, yeast replacement with a cheaper protein source could significantly decrease the cost of a mass-rearing facility. However, this replacement should not compromise the production of high biological quality insects. The yeast substitute should be of equal nutritional value, if not higher.

Following the promising results of a similar study in *C. capitata*, we tested whether *Enterobacter* sp. AA26 dry biomass could be a reliable alternative protein source. The total replacement of brewer's yeast had detrimental effects on female fecundity, pupal weight, and adult survivorship and recovery, thus indicating that *Enterobacter* sp. AA26 cannot fulfill the protein requirements of *D. suzukii* when yeast is absent from the diet. Interestingly, the partial yeast replacement did not present severe effects (apart from the adult recovery rate), suggesting that halving the yeast quantity is still sufficient to produce fit adults. In contrast, Kyritsis et al. suggested that *Enterobacter* sp. AA26 dry biomass can be used as an adequate replacement of brewer's yeast without any negative impact on the biological quality and the productivity rate of *C. capitata* [28]. The different physiology and biology of the two insect species might be one of the reasons why *Enterobacter* sp. AA26 failed to "act" as a suitable protein source. In addition, *Enterobacter* sp. AA26 is a *C. capitata* gut isolate, and although *Enterobacteriaceae* prevail in *D. suzukii* gut and *Enterobacter* sp. has been detected in a few *D. suzukii* studies, no data is available whether this specific AA26 strain is a member of the *D. suzukii* gut microbiota. Strain inconsistency or even bacterial competition could explain why *Enterobacter* sp. AA26 was an unfavorable protein source for *D. suzukii*. A recent study has shown that an infection of *E. ludwigii* in *D. melanogaster* affected the development of the flies and caused age-dependent neurodegeneration, thus indicating that specific *Enterobacter* sp. can negatively impact core aspects of the biology and behavior of the flies [85]. Further studies are required to identify promising microbe candidates with high nutritional value for *D. suzukii*.

Previous studies on fruit flies have clearly indicated that insect fitness is correlated to both the type and dose of the amino acids provided by the diet [86,87]. A study by Azis et al. 2019, compared the amino acid and vitamin content of *Enterobacter* sp. AA26 and Torula yeast [59]. *Enterobacter* sp. AA26 proved to be a sufficient source of all the essential nutrients required by *C. capitata*. Both essential and non-essential amino acids and vitamins were included in adequate amount in *Enterobacter* sp. AA26 biomass thus making it a strong candidate for the replacement of Torula yeast. Although glutamic acid and proline represented a lower fraction of the protein content compared to *Enterobacter* sp. AA26, the overall performance of the diet was equal to the Torula one for *C. capitata* larvae [59]. However, one cannot rule out the possibility that this lower percentage of glutamic acid and proline could affect the *D. suzukii* development and fitness. At the present study,

Torula yeast has not been used and therefore it is not safe to extrapolate any conclusions related to the inactive brewer's yeast used for *D. suzukii*, since the two yeast types have different nutritional properties [88].

The outcomes of our study suggest that the "lack of performance" of *Enterobacter* sp. AA26 might not be due to the protein content per se but due to the protein quality. The nutritive value of the yeast's protein content is quite high and studies on the proteinogenic composition of yeast extracts have shown that the protein proportion of brewer's yeast can be more than 60% and the proportion of free amino acids more than 30%, thus indicating a rich protein source [88,89]. Apart from the essential amino acids, the content of minerals, vitamins and lipids play a role in the nutrient composition of brewer's yeast. Significant differences between brewer's yeast and *Enterobacter* sp. AA26 in the bioavailability of nutrients, as well as in the interactions among individual nutrients could be a reason explaining the reduced performance of *Enterobacter* sp. AA26. To the best of our knowledge there is not a corresponding study that compares *Enterobacter* sp. AA26 with inactive brewer's yeast, and thus our insight regarding their differences in terms of nutritional value is limited. A future study could shed light on this aspect and reveal why *Enterobacter* sp. AA26 is not sufficient as a protein source for the development of *D. suzukii*.

5. Conclusions

In the context of an SIT application, improvement of the mass-rearing protocols is always in the center of research efforts. Our findings clearly highlighted the importance of yeast as a diet component for *D. suzukii*. The *C. capitata* gut isolate, *Enterobacter* sp. AA26, cannot replace nutritionally the inactive brewer's yeast in the *D. suzukii* larval diet. However, one should not exclude the possibility that bacterial isolates coming from the *D. suzukii* gut could have a beneficial effect on the fly productivity or be an adequate protein source that would eventually lead in yeast replacement. Future studies should focus on dissecting the gut bacterial diversity of *D. suzukii* and employ the most abundant inhabitants as candidates for the yeast replacement. In addition, cost estimation studies should also be performed for the promising candidates to elucidate their potential as core larval diet components in mass-rearing facilities.

Author Contributions: Conceptualization, K.N., C.C. and K.B.; Data curation, K.N. and F.S.; Formal analysis, K.N. and F.S.; Funding acquisition, C.S.; Investigation, K.N. and F.S.; Methodology, K.N.; Resources, S.N.; Supervision, C.S., C.C. and K.B.; Validation, K.N. and F.S.; Writing—original draft, K.N. and F.S.; Writing—review & editing, K.N., F.S., S.N., C.S., C.C. and K.B. All authors have read and agreed to the published version of the manuscript.

Funding: This study was supported by the Insect Pest Control Subprogramme of the Joint FAO/IAEA Centre of Nuclear Techniques in Food and Agriculture. This research was also funded by the Austrian Science Fund (FWF): I 2604-B25.

Institutional Review Board Statement: Not applicable.

Informed Consent Statement: Not applicable.

Data Availability Statement: The data presented in this study are available on request from the authors.

Acknowledgments: The authors would like to thank Rui Pereira and Marc Vreysen for their support throughout this study.

Conflicts of Interest: The authors declare no conflict of interest.

References

1. Bourtzis, K.; Miller, T. *Insect Symbiosis*; CRC Press, Taylor and Francis Group, LLC: Boca Raton, FL, USA, 2003.
2. Qiao, H.; Keesey, I.W.; Hansson, B.S.; Knaden, M. Gut Microbiota Affects Development and Olfactory Behavior in *Drosophila melanogaster*. *J. Exp. Biol.* **2019**, *222*, jeb.192500. [CrossRef] [PubMed]
3. Lizé, A.; McKay, R.; Lewis, Z. Kin Recognition in Drosophila: The Importance of Ecology and Gut Microbiota. *ISME J.* **2014**, *8*, 469–477. [CrossRef] [PubMed]

4.   Farine, J.-P.; Habbachi, W.; Cortot, J.; Roche, S.; Ferveur, J.-F. Maternally-Transmitted Microbiota Affects Odor Emission and Preference in *Drosophila* Larva. *Sci. Rep.* **2017**, *7*, 6062. [CrossRef] [PubMed]

5.   Douglas, A.E. Multiorganismal Insects: Diversity and Function of Resident Microorganisms. *Annu. Rev. Entomol.* **2015**, *60*, 17–34. [CrossRef]

6.   Kouser, S.; Qaim, M. Impact of Bt Cotton on Pesticide Poisoning in Smallholder Agriculture: A Panel Data Analysis. *Ecol. Econ.* **2011**, *70*, 2105–2113. [CrossRef]

7.   Crotti, E.; Balloi, A.; Hamdi, C.; Sansonno, L.; Marzorati, M.; Gonella, E.; Favia, G.; Cherif, A.; Bandi, C.; Alma, A.; et al. Microbial Symbionts: A Resource for the Management of Insect-Related Problems: Microbial Resource Management Applied to Insects. *Microb. Biotechnol.* **2012**, *5*, 307–317. [CrossRef]

8.   Deutscher, A.T.; Reynolds, O.L.; Chapman, T.A. Yeast: An Overlooked Component of *Bactrocera tryoni* (Diptera: Tephritidae) Larval Gut Microbiota. *J. Econ. Entomol.* **2016**, *110*, 298–300. [CrossRef] [PubMed]

9.   Yuval, B.; Ben-Ami, E.; Behar, A.; Ben-Yosef, M.; Jurkevitch, E. The Mediterranean Fruit Fly and Its Bacteria—Potential for Improving Sterile Insect Technique Operations: Bacteria of *Ceratitis capitata* and the SIT. *J. Appl. Entomol.* **2013**, *137*, 39–42. [CrossRef]

10.   Dyck, V.A.; Hendrichs, J.; Robinson, A.S. *Sterile Insect Technique: Principles and Practice in Area-Wide Integrated Pest Management*, 2nd ed.; Dyck, V.A., Hendrichs, J., Robinson, A.S., Eds.; CRC Press: Boca Raton, FL, USA, 2021; ISBN 978-1-00-303557-2.

11.   Enkerlin, W.R. Impact of Fruit Fly Control Programmes Using the Sterile Insect Technique. In *Sterile Insect Technique: Principles and Practice in Area-Wide Integrated Pest Management*, 2nd ed.; Dyck, V.A., Hendrichs, J., Robinson, A.S., Eds.; CRC Press: Boca Raton, FL, USA, 2021; ISBN 978-1-00-303557-2.

12.   Knipling, E.F. Possibilities of Insect Control or Eradication Through the Use of Sexually Sterile Males1. *J. Econ. Entomol.* **1955**, *48*, 459–462. [CrossRef]

13.   Parker, A.G. Mass-Rearing for the Sterile Insect Technique. In *Sterile Insect Technique: Principles and Practice in Area-Wide Integrated Pest Management*, 2nd ed.; Dyck, V.A., Hendrichs, J., Robinson, A.S., Eds.; CRC Press: Boca Raton, FL, USA, 2021; ISBN 978-1-00-303557-2.

14.   Chaudhury, M.F.B. *Handbook of Insect Rearing, Vol. I and II*; Singh, P., Moore, R.F., Eds.; Elsevier: Amsterdam, The Netherlands, 1985.

15.   Cohen, A.C. Ecology of Insect Rearing Systems: A Mini-Review of Insect Rearing Papers from 1906-2017. *Adv. Entomol.* **2018**, *6*, 86–115. [CrossRef]

16.   Shimoji, Y.; Miyatake, T. Adaptation to Artificial Rearing During Successive Generations in the West Indian Sweetpotato Weevil, *Euscepes postfasciatus* (Coleoptera: Curculionidae). *Ann. Entomol. Soc. Am.* **2002**, *95*, 735–739. [CrossRef]

17.   Niyazi, N.; Lauzon, C.R.; Shelly, T.E. Effect of Probiotic Adult Diets on Fitness Components of Sterile Male Mediterranean Fruit Flies (Diptera: Tephritidae) Under Laboratory and Field Cage Conditions. *J. Econ. Entomol.* **2004**, *97*, 1570–1580. [CrossRef]

18.   Ami, E.B.; Yuval, B.; Jurkevitch, E. Manipulation of the Microbiota of Mass-Reared Mediterranean Fruit Flies *Ceratitis capitata* (Diptera: Tephritidae) Improves Sterile Male Sexual Performance. *ISME J.* **2010**, *4*, 28–37. [CrossRef] [PubMed]

19.   Rull, J.; Lasa, R.; Rodriguez, C.; Ortega, R.; Velazquez, O.E.; Aluja, M. Artificial Selection, Pre-Release Diet, and Gut Symbiont Inoculation Effects on Sterile Male Longevity for Area-Wide Fruit-Fly Management. *Entomol. Exp. Appl.* **2015**, *157*, 325–333. [CrossRef]

20.   Meats, A.; Streamer, K.; Gilchrist, A.S. Bacteria as Food Had No Effect on Fecundity during Domestication of the Fruit Fly, *Bactrocera tryoni*. *J. Appl. Entomol.* **2009**, *133*, 633–639. [CrossRef]

21.   Gavriel, S.; Jurkevitch, E.; Gazit, Y.; Yuval, B. Bacterially Enriched Diet Improves Sexual Performance of Sterile Male Mediterranean Fruit Flies: Bacteria and Medfly Sexual Performance. *J. Appl. Entomol.* **2011**, *135*, 564–573. [CrossRef]

22.   Hamden, H.; M'saad Guerfali, M.; Fadhl, S.; Saidi, M.; Chevrier, C. Fitness Improvement of Mass-Reared Sterile Males of *Ceratitis capitata* (Vienna 8 Strain) (Diptera: Tephritidae) After Gut Enrichment With Probiotics. *J. Econ. Entomol.* **2013**, *106*, 641–647. [CrossRef] [PubMed]

23.   Augustinos, A.A.; Kyritsis, G.A.; Papadopoulos, N.T.; Abd-Alla, A.M.M.; Cáceres, C.; Bourtzis, K. Exploitation of the Medfly Gut Microbiota for the Enhancement of Sterile Insect Technique: Use of *Enterobacter* sp. in Larval Diet-Based Probiotic Applications. *PLoS ONE* **2015**, *10*, e0136459. [CrossRef]

24.   Kyritsis, G.A.; Augustinos, A.A.; Cáceres, C.; Bourtzis, K. Medfly Gut Microbiota and Enhancement of the Sterile Insect Technique: Similarities and Differences of *Klebsiella oxytoca* and *Enterobacter* sp. AA26 Probiotics during the Larval and Adult Stages of the VIENNA 8D53+ Genetic Sexing Strain. *Front. Microbiol.* **2017**, *8*, 2064. [CrossRef]

25.   Cai, Z.; Yao, Z.; Li, Y.; Xi, Z.; Bourtzis, K.; Zhao, Z.; Bai, S.; Zhang, H. Intestinal Probiotics Restore the Ecological Fitness Decline of *Bactrocera dorsalis* by Irradiation. *Evol. Appl.* **2018**, *11*, 1946–1963. [CrossRef]

26.   Khaeso, K.; Andongma, A.A.; Akami, M.; Souliyanonh, B.; Zhu, J.; Krutmuang, P.; Niu, C.-Y. Assessing the Effects of Gut Bacteria Manipulation on the Development of the Oriental Fruit Fly, *Bactrocera dorsalis* (Diptera; Tephritidae). *Symbiosis* **2018**, *74*, 97–105. [CrossRef]

27.   Sacchetti, P.; Ghiardi, B.; Granchietti, A.; Stefanini, F.M.; Belcari, A. Development of Probiotic Diets for the Olive Fly: Evaluation of Their Effects on Fly Longevity and Fecundity. *Ann. Appl. Biol.* **2014**, *164*, 138–150. [CrossRef]

28. Kyritsis, G.A.; Augustinos, A.A.; Ntougias, S.; Papadopoulos, N.T.; Bourtzis, K.; Cáceres, C. *Enterobacter* sp. AA26 Gut Symbiont as a Protein Source for Mediterranean Fruit Fly Mass-Rearing and Sterile Insect Technique Applications. *BMC Microbiol.* **2019**, *19*, 288. [CrossRef]

29. Ben-Yosef, M.; Pasternak, Z.; Jurkevitch, E.; Yuval, B. Symbiotic Bacteria Enable Olive Fly Larvae to Overcome Host Defences. *R. Soc. Open Sci.* **2015**, *2*, 150170. [CrossRef] [PubMed]

30. Capuzzo, C.; Firrao, G.; Mazzon, L.; Squartini, A.; Girolami, V. *Candidatus Erwinia dacicola*, a Coevolved Symbiotic Bacterium of the Olive Fly Bactrocera Oleae (Gmelin). *Int. J. Syst. Evol. Microbiol.* **2005**, *55*, 1641–1647. [CrossRef]

31. Hauser, M. A Historic Account of the Invasion of *Drosophila suzukii* (Matsumura) (Diptera: Drosophilidae) in the Continental United States, with Remarks on Their Identification. *Pest. Manag. Sci.* **2011**, *67*, 1352–1357. [CrossRef] [PubMed]

32. dos Santos, L.A.; Mendes, M.F.; Krüger, A.P.; Blauth, M.L.; Gottschalk, M.S.; Garcia, F.R.M. Global Potential Distribution of *Drosophila suzukii* (Diptera: Drosophilidae). *PLoS ONE.* **2017**, *12*, e0174318. [CrossRef] [PubMed]

33. Calabria, G.; Máca, J.; Bächli, G.; Serra, L.; Pascual, M. First Records of the Potential Pest Species *Drosophila suzukii* (Diptera: Drosophilidae) in Europe: First Record of *Drosophila suzukii* in Europe. *J. Appl. Entomol.* **2012**, *136*, 139–147. [CrossRef]

34. Deprá, M.; Poppe, J.L.; Schmitz, H.J.; De Toni, D.C.; Valente, V.L.S. The First Records of the Invasive Pest *Drosophila suzukii* in the South American Continent. *J. Pest. Sci.* **2014**, *87*, 379–383. [CrossRef]

35. Bellamy, D.E.; Sisterson, M.S.; Walse, S.S. Quantifying Host Potentials: Indexing Postharvest Fresh Fruits for Spotted Wing Drosophila, *Drosophila suzukii*. *PLoS ONE* **2013**, *8*, e61227. [CrossRef]

36. Mitsui, H.; Takahashi, K.H.; Kimura, M.T. Spatial Distributions and Clutch Sizes of *Drosophila* Species Ovipositing on Cherry Fruits of Different Stages. *Popul. Ecol.* **2006**, *48*, 233–237. [CrossRef]

37. De Ros, G.; Grassi, A.; Pantezzi, T. Recent Trends in the Economic Impact of *Drosophila suzukii*. In *Drosophila suzukii Management*; Garcia, F.R.M., Ed.; Springer International Publishing: Cham, Switzerland, 2020; pp. 11–27. ISBN 978-3-030-62691-4.

38. Lee, J.C.; Wang, X.; Daane, K.M.; Hoelmer, K.A.; Isaacs, R.; Sial, A.A.; Walton, V.M. Biological Control of Spotted-Wing Drosophila (Diptera: Drosophilidae)—Current and Pending Tactics. *J. Integr. Pest. Manag.* **2019**, *10*, 13. [CrossRef]

39. Leach, H.; Van Timmeren, S.; Isaacs, R. Exclusion Netting Delays and Reduces *Drosophila suzukii* (Diptera: Drosophilidae) Infestation in Raspberries. *J. Econ. Entomol.* **2016**, *109*, 2151–2158. [CrossRef] [PubMed]

40. Nikolouli, K.; Colinet, H.; Renault, D.; Enriquez, T.; Mouton, L.; Gibert, P.; Sassu, F.; Cáceres, C.; Stauffer, C.; Pereira, R.; et al. Sterile Insect Technique and *Wolbachia* Symbiosis as Potential Tools for the Control of the Invasive Species *Drosophila suzukii*. *J. Pest. Sci.* **2018**, *91*, 489–503. [CrossRef]

41. Nikolouli, K.; Sassù, F.; Mouton, L.; Stauffer, C.; Bourtzis, K. Combining Sterile and Incompatible Insect Techniques for the Population Suppression of *Drosophila suzukii*. *J. Pest. Sci.* **2020**, *93*, 647–661. [CrossRef]

42. Sassù, F.; Nikolouli, K.; Stauffer, C.; Bourtzis, K.; Caceres, C. Sterile insect technique and incompatible insect technique for the integrated *Drosophila suzukii* management. In *Drosophila suzukii Management*; Garcia, F.R.M., Ed.; Springer: Cham, Switzerland. [CrossRef]

43. Sassù, F.; Nikolouli, K.; Pereira, R.; Vreysen, M.J.B.; Stauffer, C.; Cáceres, C. Irradiation Dose Response under Hypoxia for the Application of the Sterile Insect Technique in *Drosophila suzukii*. *PLoS ONE* **2019**, *14*, e0226582. [CrossRef]

44. Sassù, F.; Nikolouli, K.; Caravantes, S.; Taret, G.; Pereira, R.; Vreysen, M.J.B.; Stauffer, C.; Cáceres, C. Mass-Rearing of *Drosophila suzukii* for Sterile Insect Technique Application: Evaluation of Two Oviposition Systems. *Insects* **2019**, *10*, 448. [CrossRef]

45. Hamby, K.A.; Becher, P.G. Current Knowledge of Interactions between *Drosophila suzukii* and Microbes, and Their Potential Utility for Pest Management. *J. Pest. Sci.* **2016**, *89*, 621–630. [CrossRef]

46. Fountain, M.T.; Bennett, J.; Cobo-Medina, M.; Conde Ruiz, R.; Deakin, G.; Delgado, A.; Harrison, R.; Harrison, N. Alimentary Microbes of Winter-Form *Drosophila suzukii*. *Insect. Mol. Biol.* **2018**, *27*, 383–392. [CrossRef]

47. Vacchini, V.; Gonella, E.; Crotti, E.; Prosdocimi, E.M.; Mazzetto, F.; Chouaia, B.; Callegari, M.; Mapelli, F.; Mandrioli, M.; Alma, A.; et al. Bacterial Diversity Shift Determined by Different Diets in the Gut of the Spotted Wing Fly *Drosophila suzukii* Is Primarily Reflected on Acetic Acid Bacteria: Acetic Acid Bacteria of *Drosophila suzukii*. *Environ. Microbiol. Rep.* **2017**, *9*, 91–103. [CrossRef]

48. Jiménez-Padilla, Y.; Esan, E.O.; Floate, K.D.; Sinclair, B.J. Persistence of Diet Effects on the Microbiota of *Drosophila suzukii* (Diptera: Drosophilidae). *Can. Entomol.* **2020**, *152*, 516–531. [CrossRef]

49. Bing, X.; Gerlach, J.; Loeb, G.; Buchon, N. Nutrient-Dependent Impact of Microbes on *Drosophila suzukii* Development. *mBio* **2018**, *9*, e02199-17. [CrossRef]

50. Chandler, J.A.; James, P.M.; Jospin, G.; Lang, J.M. The Bacterial Communities of *Drosophila suzukii* Collected from Undamaged Cherries. *PeerJ* **2014**, *2*, e474. [CrossRef]

51. Chandler, J.A.; Morgan Lang, J.; Bhatnagar, S.; Eisen, J.A.; Kopp, A. Bacterial Communities of Diverse *Drosophila* Species: Ecological Context of a Host–Microbe Model System. *PLoS Genet.* **2011**, *7*, e1002272. [CrossRef]

52. Solomon, G.M.; Dodangoda, H.; McCarthy-Walker, T.; Ntim-Gyakari, R.; Newell, P.D. The Microbiota of *Drosophila suzukii* Influences the Larval Development of *Drosophila melanogaster*. *PeerJ* **2019**, *7*, e8097. [CrossRef]

53. Martinez-Sañudo, I.; Simonato, M.; Squartini, A.; Mori, N.; Marri, L.; Mazzon, L. Metagenomic Analysis Reveals Changes of the *Drosophila suzukii* Microbiota in the Newly Colonized Regions: Microbiota Associated to *Drosophila suzukii*. *Insect Sci.* **2018**, *25*, 833–846. [CrossRef]

54. Brummel, T.; Ching, A.; Seroude, L.; Simon, A.F.; Benzer, S. Drosophila Lifespan Enhancement by Exogenous Bacteria. *Proc. Natl. Acad. Sci. USA* **2004**, *101*, 12974–12979. [CrossRef]

55. Cox, C.R.; Gilmore, M.S. Native Microbial Colonization of *Drosophila melanogaster* and Its Use as a Model of *Enterococcus faecalis* Pathogenesis. *Infect. Immun.* **2007**, *75*, 1565–1576. [CrossRef]

56. Hiebert, N.; Carrau, T.; Bartling, M.; Vilcinskas, A.; Lee, K.-Z. Identification of Entomopathogenic Bacteria Associated with the Invasive Pest *Drosophila suzukii* in Infested Areas of Germany. *J. Invertebr. Pathol.* **2020**, *173*, 107389. [CrossRef]

57. Shu, R.; Hahn, D.A.; Jurkevitch, E.; Liburd, O.E.; Yuval, B.; Wong, A.C.-N. Sex-Dependent Effects of the Microbiome on Foraging and Locomotion in *Drosophila suzukii*. *Front. Microbiol.* **2021**, *12*, 656406. [CrossRef]

58. Keebaugh, E.S.; Yamada, R.; Obadia, B.; Ludington, W.B.; Ja, W.W. Microbial Quantity Impacts Drosophila Nutrition, Development, and Lifespan. *iScience* **2018**, *4*, 247–259. [CrossRef]

59. Azis, K.; Zerva, I.; Melidis, P.; Caceres, C.; Bourtzis, K.; Ntougias, S. Biochemical and Nutritional Characterization of the Medfly Gut Symbiont *Enterobacter* sp. AA26 for Its Use as Probiotics in Sterile Insect Technique Applications. *BMC Biotechnol.* **2019**, *19*, 90. [CrossRef]

60. FAO/IAEA/USDA. Product Quality Control for Sterile Mass-Reared and Released Tephritid Fruit Flies. 2019. Available online: https://www.iaea.org/sites/default/files/qcv7.pdf (accessed on 17 September 2021).

61. R Core Team. R: A Language and Environment for Statistical Computing. 2021. Available online: https://www.R-project.org/ (accessed on 17 September 2021).

62. Demétrio, C.G.B.; Hinde, J.; Moral, R.A. Models for Overdispersed Data in Entomology. In *Ecological Modelling Applied to Entomology*; Ferreira, C.P., Godoy, W.A.C., Eds.; Springer International Publishing: Cham, Switzerland, 2014; pp. 219–259. ISBN 978-3-319-06876-3.

63. Ver Hoef, J.M.; Boveng, P.L. Quasi-Poisson vs. Negative Binomial Regression: How Should We Model Overdispersed Count Data? *Ecology* **2007**, *88*, 2766–2772. [CrossRef]

64. Nelder, J.A.; Wedderburn, R.W.M. Generalized Linear Models. *J. R. Stat. Soc. Ser. A. Stat. Soc.* **1972**, *135*, 370. [CrossRef]

65. Dunn, P.K.; Smyth, G.K. *Generalized Linear Models With Examples in R.*, 1st ed.; Springer: New York, NY, USA, 2018; ISBN 978-1-4419-0118-7.

66. Kaplan, E.L.; Meier, P. Nonparametric Estimation from Incomplete Observations. *J. Am. Stat. Assoc.* **1958**, *53*, 457–481. [CrossRef]

67. Therneau, T.M.; Grambsch, P.M. *Modeling Survival Data: Extending the Cox Model*; Springer: New York, NY, USA, 2000; ISBN 978-0-387-98784-2.

68. Moral, R.A.; Hinde, J.; Demétrio, C.G.B. Half-Normal Plots and Overdispersed Models in R: The Hnp Package. *J. Stat. Soft.* **2017**, *81*(10), 1–23. [CrossRef]

69. Hartig, F. DHARMa: Residual Diagnostics for Hierarchical (Multi-Level/Mixed) Regression Models. 2021. R Package Version. Available online: http://florianhartig.github.io/DHARMa/ (accessed on 17 September 2021).

70. Lenth, R.V. Least-Squares Means: The *R* Package Lsmeans. *J. Stat. Soft.* **2016**, *69*, 1–33. [CrossRef]

71. Lesperance, D.N.; Broderick, N.A. Microbiomes as Modulators of *Drosophila melanogaster* Homeostasis and Disease. *Curr. Opin. Insect Sci.* **2020**, *39*, 84–90. [CrossRef]

72. Corby-Harris, V.; Pontaroli, A.C.; Shimkets, L.J.; Bennetzen, J.L.; Habel, K.E.; Promislow, D.E.L. Geographical Distribution and Diversity of Bacteria Associated with Natural Populations of *Drosophila melanogaster*. *Appl. Environ. Microbiol.* **2007**, *73*, 3470–3479. [CrossRef]

73. Broderick, N.A.; Lemaitre, B. Gut-Associated Microbes of *Drosophila melanogaster*. *Gut Microbes* **2012**, *3*, 307–321. [CrossRef]

74. Augustinos, A.A. Insect Symbiosis in Support of the Sterile Insect Technique. In *Sterile Insect Technique: Principles and Practice in Area-Wide Integrated Pest Management*, 2nd ed.; Dyck, V.A., Hendrichs, J., Robinson, A.S., Eds.; CRC Press: Boca Raton, FL, USA, 2021; ISBN 978-1-00-303557-2.

75. Geiger, A.; Fardeau, M.-L.; Grebaut, P.; Vatunga, G.; Josénando, T.; Herder, S.; Cuny, G.; Truc, P.; Ollivier, B. First Isolation of *Enterobacter*, *Enterococcus*, and *Acinetobacter* spp. as Inhabitants of the Tsetse Fly (*Glossina palpalis palpalis*) Midgut. *Infect. Genet. Evol.* **2009**, *9*, 1364–1370. [CrossRef] [PubMed]

76. Jiang, J.; Alvarez, C.; Kukutla, P.; Yu, W.; Xu, J. Draft Genome Sequences of *Enterobacter* sp. Isolate Ag1 from the Midgut of the Malaria Mosquito Anopheles Gambiae. *J. Bacteriol.* **2012**, *194*, 5481. [CrossRef] [PubMed]

77. Gujjar, N.R.; Govindan, S.; Verghese, A.; Subramaniam, S.; More, R. Diversity of the Cultivable Gut Bacterial Communities Associated with the Fruit Flies *Bactrocera dorsalis* and *Bactrocera cucurbitae* (Diptera: Tephritidae). *Phytoparasitica* **2017**, *45*, 453–460. [CrossRef]

78. Hardin, J.A.; Kraus, D.A.; Burrack, H.J. Diet Quality Mitigates Intraspecific Larval Competition in *Drosophila suzukii*. *Entomol. Exp. Appl.* **2015**, *156*, 59–65. [CrossRef]

79. Rendon, D.; Walton, V.; Tait, G.; Buser, J.; Lemos Souza, I.; Wallingford, A.; Loeb, G.; Lee, J. Interactions among Morphotype, Nutrition, and Temperature Impact Fitness of an Invasive Fly. *Ecol. Evol.* **2019**, *9*, 2615–2628. [CrossRef]

80. Lee, J.C.; Burrack, H.J.; Barrantes, L.D.; Beers, E.H.; Dreves, A.J.; Hamby, K.A.; Haviland, D.R.; Isaacs, R.; Richardson, T.A.; Shearer, P.W.; et al. Evaluation of Monitoring Traps for *Drosophila suzukii* (Diptera: Drosophilidae) in North America. *J. Econ. Entomol.* **2012**, *105*, 1350–1357. [CrossRef]

81. Lewis, M.T.; Hamby, K.A. Differential Impacts of Yeasts on Feeding Behavior and Development in Larval *Drosophila suzukii* (Diptera:Drosophilidae). *Sci. Rep.* **2019**, *9*, 13370. [CrossRef]

82. Bellutti, N.; Gallmetzer, A.; Innerebner, G.; Schmidt, S.; Zelger, R.; Koschier, E.H. Dietary Yeast Affects Preference and Performance in *Drosophila suzukii*. *J. Pest. Sci.* **2018**, *91*, 651–660. [CrossRef]

83. Silva-Soares, N.F.; Nogueira-Alves, A.; Beldade, P.; Mirth, C.K. Adaptation to New Nutritional Environments: Larval Performance, Foraging Decisions, and Adult Oviposition Choices in *Drosophila suzukii*. *BMC Ecol.* **2017**, *17*, 21. [CrossRef]
84. Young, Y.; Buckiewicz, N.; Long, T.A.F. Nutritional Geometry and Fitness Consequences in *Drosophila suzukii*, the Spotted-Wing Drosophila. *Ecol. Evol.* **2018**, *8*, 2842–2851. [CrossRef]
85. Priyadarsini, S.; Sahoo, M.; Sahu, S.; Jayabalan, R.; Mishra, M. An Infection of *Enterobacter ludwigii* Affects Development and Causes Age-Dependent Neurodegeneration in *Drosophila melanogaster*. *Invert. Neurosci.* **2019**, *19*, 13. [CrossRef] [PubMed]
86. Zografou, E.N.; Tsiropoulos, G.J.; Margaritis, L.H. Survival, Fecundity and Fertility of *Bactrocera oleae*, as Affected by Amino Acid Analogues. *Entomol. Exp. Appl.* **1998**, *87*, 125–132. [CrossRef]
87. Chang, C.L.; Albrecht, C.; El-Shall, S.S.A.; Kurashima, R. Adult Reproductive Capacity of *Ceratitis capitata* (Diptera: Tephritidae) on a Chemically Defined Diet. *Ann. Entomol. Soc. Am.* **2001**, *94*, 702–706. [CrossRef]
88. Jacob, F.F.; Michel, M.; Zarnkow, M.; Hutzler, M.; Methner, F.-J. The Complexity of Yeast Extracts and Its Consequences on the Utility in Brewing: A Review. *Brew. Sci.* **2019**, 50–62. [CrossRef]
89. Yamada, E.A.; Sgarbieri, V.C. Yeast (Saccharomyces cerevisiae) Protein Concentrate: Preparation, Chemical Composition, and Nutritional and Functional Properties. *J. Agric. Food Chem.* **2005**, *53*, 3931–3936. [CrossRef]

insects

Article

Effect of Cold Storage on the Quality of *Psyttalia incisi* (Hymenoptera: Braconidae), a Larval Parasitoid of *Bactrocera dorsalis* (Diptera: Tephritidae)

Jia Lin [1,2,3], Deqing Yang [1,2,3], Xuxing Hao [1,2,3], Pumo Cai [1,2,3,4], Yaqing Guo [1,2,3], Shuang Shi [1,2,3], Changming Liu [1,2,3] and Qinge Ji [1,2,3,*]

[1] Institute of Beneficial Insects, Plant Protection College, Fujian Agriculture and Forestry University, Fuzhou 350002, China; Linjia@fafu.edu.cn (J.L.); 3200231054@fafu.edu.cn (D.Y.); 1200203004@fafu.edu.cn (X.H.); caipumo@fafu.edu.cn (P.C.); 3190231007@fafu.edu.cn (Y.G.); 1190203017@fafu.edu.cn (S.S.); cmliu@fjau.edu.cn (C.L.)
[2] Key Laboratory of Biopesticide and Chemical Biology, Ministry of Education, Fuzhou 350002, China
[3] State Key Laboratory of Ecological Pest Control for Fujian and Taiwan Crops, Fuzhou 350002, China
[4] Department of Horticulture, College of Tea and Food Science, Wuyi University, Wuyishan 354300, China
* Correspondence: jiqinge@fafu.edu.cn

Simple Summary: Biological control programs primarily rely on the mass-release of high-quality bioagents in order to successfully suppress pests. However, producing such bioagents on a large scale and within a short timeframe or in a single step is extremely difficult. Therefore, it is important to consider methods that could increase the shelf life and help to synchronize the release schedule of bioagents reared in different batches. In the present study, we determined the effects of various cold storage protocols on the emergence and quality of *Psyttalia incisi*, a larval parasitoid of *Bactrocera dorsalis*. Our results indicated that there were no negative impacts on the emergence parameters and adult quality when late-age *P. incisi* pupae were stored at 13 °C for 10 or 15 d. This information is valuable in facilitating the mass-rearing of *P. incisi* and helping to improve the efficiency of biological control programs using *P. incisi* against *B. dorsalis*.

Abstract: *Psyttalia incisi* (Silvestri) is the dominant parasitoid against *Bactrocera dorsalis* (Hendel) in fruit-producing regions of southern China. Prior to a large-scale release, it is important to generate a sufficient stockpile of *P. incisi* whilst considering how best to maintain their quality and performance; cold storage is an ideal method to achieve these aims. In this study, the impacts of temperature and storage duration on the developmental parameters of *P. incisi* pupae at different age intervals were assessed. Then, four of the cold storage protocols were chosen for further evaluating their impacts on the quality parameters of post-storage adults. Results showed that the emergence rate of *P. incisi* was significantly affected by storage temperature, storage duration, and pupal age interval and their interactions. However, when late-age *P. incisi* pupae developed at a temperature of 13 °C for 10 or 15 d, no undesirable impacts on dry weight, flight ability, longevity, reproduction parameters of post-storage adults, emergence rate, or the female proportion of progeny were recorded. Our findings demonstrate that cold storage has the potential for enhancing the flexibility and effectiveness of the large-scale production and application of *P. incisi*.

Keywords: *Psyttalia incisi*; oriental fruit fly; cold storage; emergence rate; quality; reproduction

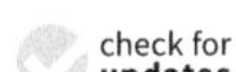

Citation: Lin, J.; Yang, D.; Hao, X.; Cai, P.; Guo, Y.; Shi, S.; Liu, C.; Ji, Q. Effect of Cold Storage on the Quality of *Psyttalia incisi* (Hymenoptera: Braconidae), a Larval Parasitoid of *Bactrocera dorsalis* (Diptera: Tephritidae). *Insects* **2021**, *12*, 558. https://doi.org/10.3390/insects 12060558

Academic Editor: Man P. Huynh

Received: 11 May 2021
Accepted: 4 June 2021
Published: 16 June 2021

Publisher's Note: MDPI stays neutral with regard to jurisdictional claims in published maps and institutional affiliations.

1. Introduction

Batrocera dorsalis (Hendel) (Diptera: Tephritidae) is a notorious pest of economic importance, largely due to its traits of polyphagia, superior dispersal ability, outstanding climate adaptability, and high fecundity [1]. Over the last two decades, this fly has spread into many tropical and subtropical regions due to both human transportation and adult fly migration, causing considerable damage to commercial fruits and horticultural products as well as the associated import and export trade [2,3]. At present, the primary strategy for suppressing this pest involves spraying chemical insecticides, either alone or in combination with food-based lures [4]. However, there are numerous problems associated with this strategy, including insecticide resistance, environmental depredation, and effects on food safety, which has led to pressure for an alternative strategy to be developed [5,6]. Biological control, which has the advantages of being long-lasting and environmentally friendly, has recently gained much attention and is regarded as the prime alternative tactic against *B. dorsalis* populations [7].

Psyttalis incisi (Silvestri) (Hymenoptera: Braconidae: Opiinae) is a solitary opiine endoparasitoid, whose preferential host is the early larval instar of *B. dorsalis* [8]. It has been reported that *P. incisi* exist naturally and occupy a dominant proportion (77.6%) of the parasitic wasp population against *B. dorsalis* in fields of Zhangzhou City, Fujian Province, China, and contributed a limited reduction in *B. dorsalis* populations [9]. Hence, *P. incisi* is a highly suitable bioagent for biological control programs against *B. dorsalis* in that region of China. However, in order for such programs to be effective, millions of parasitoids need to be first produced and then transported into the affected areas. In order to improve the ease at which this can be achieved, it is important to consider methods by which the shelf life of *P. incisi* can be increased. This will help to ensure a sufficient stockpile of parasitoids and allow releases to be appropriately timed given their physiological characteristics.

Exposing bioagents to low temperatures can extend their developmental duration and is particularly valuable for the inundative biological control program which entailed a large number of biocontrol agents [10]. However, keeping parasitoids at a sub-ambient temperature can result in cold injuries and excessive consumption of energy reserves, resulting in undesirable effects on the quality and quantity of post-storage parasitoids [11]. Hence, to minimize losses in the performance of parasitoids after undergoing storage, packaging, and shipment, recent research has focused on optimizing the tradeoffs between quality and the cold storage protocol [12–14]. For braconid parasitoids against tephritid pests, the impacts of cold storage have been evaluated in several species, such as *Psyttalia humilis* (Silvestri), *Psyttalia ponerophaga* (Silvestri), *Fopius arisanus* (Sonan), and *Diachasmimorpha longicaudata* (Ashmead) (Hymenoptera: Braconidae); these works have provided a lot of valuable information for facilitating the practical application of biological control programs [15–17]. However, to date, cold storage has never been investigated as an auxiliary approach for the large-scale rearing and release of *P. incisi*. Owing to the vital role of *P. incisi* in the management of *B. doralis*, it is crucial to optimize the flexibility and effectiveness of large-scale production and mass-release programs of *P. incisi* through cold storage techniques.

In the present study, we first determined the effects of various storage temperatures (4, 7, 10, and 13 °C) and exposure durations (10, 15, 20, and 25 days) on the emergence rate of *P. incisi* at different pupal age intervals within parasitized *B. dorsalis* puparia. Further experiments were then carried out to examine the dry weight, flight ability, longevity, and reproduction parameters of parental *P. incisi* (G1) along with emergence rates, and the female proportion of progeny (G2).

2. Materials and Methods

2.1. Insect Colonies

Bactrocera dorsalis and *P. incisi* were initially collected from orchards in and around Fujian institutes of Tropical Crops (117°30′58.95″ E, 24°37′31.37″ N, and 26 m altitude), in Zhangzhou City, Fujian Province, China in 2004. These orchards are composed of 'Pearl' guava (*Psidium guajava* L.), carambola (*Averrhoa carambola* L.), and wax apple (*Syzygium samarangense*), the annual average temperature of this area was 21 °C and annual average rainfall was 1603 mm [9]. *Bactrocera dorsalis* and *P. incisi* were then identified in the laboratory [18], and these vouchers were deposited in the UN (China) Center for Fruit Fly Prevention and Treatment, Fujian Agriculture and Forestry University. *Bactrocera dorsalis* were permitted to oviposit in a plastic bottle that was neatly pierced with holes; eggs were collected and transferred to a tray containing a mill feed diet for larval development, prepared in accordance with Chang et al. [19]. Puparia were collected from the bottom of the tray and transferred into a gauze cage ($30 \times 30 \times 30$ cm^3) until emergence. Fly adults were provided with a diet of yeast extract and sugar (1:3, wt:wt) and water. For the rearing of *P. incisi*, excessive numbers of second-instar *B. dorsalis* larvae were transferred to an oviposition plate (diameter: 9 cm, high: 0.5 cm) which was covered with 80 mesh net to prevent larvae from escaping. Then, two oviposition plates were provided to 500 pairs of *P. incisi* adults within a cage for 24 h to avoid super-parasitism. Honey and water were provided for *P. incisi* adults. *Bactrocera dorsalis* and *P. incisi* used in the experiments were reared under the controlled conditions of 25 ± 1 °C, $65 \pm 5\%$ relative humidity (RH), and a 12:12 h (L:D) photoperiod.

2.2. Effects of Storage Temperature, Storage Duration and Pupal Age Interval on the Emergence Parameters of P. incisi

Newly-formed *B. dorsalis* puparia were collected daily, with parasitized and unparasitized puparia distinguished by several characteristics as described in Wang and Messing [20] and Danne et al. [15]. Briefly, for parasitized puparia, the major characteristics included oviposition scars on the cuticle, a gap inside the fly pupa caused by consumption of the fly body by a parasitoid larva, and being relatively smaller and browner. For unparasitized puparia, the appendages of the fly could be clearly observed through the cuticle under a microscope.

Based on a preliminary experiment that identified the immature stage of *P. incisi* (through dissecting puparia), parasitized *B. dorsalis* puparia were incubated at 25 °C for 3, 6, and 9 days for *P. incisi* to develop into prepupae, middle-age pupae (some appendages, the shape of abdomen and thorax are observable but without any color; eye and ocelli are rufous), and late-age pupae (body is tawny, eye and ocelli are dark, mouth is dark brown), respectively. The three developmental stages were subsequently stored in four incubators (PRX-25013, Safu, Ningbo, China) set at constant low temperatures (4, 7, 10, and 13 °C, respectively) for four durations (10, 15, 20, and 25 d), all at 75% RH. A control group comprised parasitized puparia that developed in an incubator set at 25 °C and 75% RH until emergence. For each treatment and control, Petri dishes contained 30 parasitized puparia each were prepared. To ensure a ventilated condition to avoid the outbreak of pathogens, gauze was used to cover the dishes. Dishes were randomly assigned to the various treatment groups and inspections of each incubator were performed daily (for 30 s) to observe whether *P. incisi* adults had emerged during the cold exposure period. After cold exposure, treatments were held under control conditions and the number and sex of emerged *P. incisi* adults were recorded daily. All unemerged puparia of each treatment were dissected 5 days after the last *P. incisi* emerged or 15 days after cold treatment protocol (for those treatments without *P. incisi* emerged). A total of nine replicates were performed for this experiment.

2.3. Effects of Pupal Cold Storage on Quality of P. incisi Adults

To determine the effects of cold exposure on the G1 quality and G2 emergence parameters, four pupal cold storage protocols that did not have significant negative impacts on the emergence rate of *P. incisi* were selected for further experiments based on the results of Experiment 2.2. After the cold storage protocol, this series of bioassays were conducted under controlled conditions of 25 ± 1 °C and $65 \pm 5\%$ RH.

The four cold storage protocols were as follows:

Cold storage 1 (CS1): late-age *P. incisi* pupa (9-day-old parasitized *B. dorsalis* puparia) stored at 13 °C for 10 days;

Cold storage 2 (CS2): middle-age *P. incisi* pupa (6-day-old parasitized *B. dorsalis* puparia) stored at 13 °C for 10 days;

Cold storage 3 (CS3): late-age *P. incisi* pupa (9-day-old parasitized *B. dorsalis* puparia) stored at 13 °C for 15 days;

Cold storage 4 (CS4): middle-age *P. incisi* pupa (6-day-old parasitized *B. dorsalis* puparia) stored at 13 °C for 15 days.

2.3.1. Dry Weight

Parasitized puparia were transferred to a plastic bowl within a gauze cage. Parasitoids that had emerged within 24 h without foraging any food were used in this bioassay. For each treatment, 1 group of 50 females and 1 group of 50 males were respectively placed inside a total of 6 centrifuge tubes. All adults were kept at –20 °C for 20 min and were then dried in an oven at 60 °C for 48 h. Subsequently, the gross dry weight of a cohort of 50 females or males from each group was measured using a semimicro balance (CP225D, accuracy of 0.01 mg, Sartorius, Göttingen, Germany). Nine replicates were conducted in this experiment.

2.3.2. Flight Capacity

Fifty newly-emerged (<24 h) females and 50 newly-emerged males were caught using 5 mL centrifuge tubes. These centrifuge tubes were held at 4 °C for 10 s to immobilize the adult wasps. Subsequently, two tubes each of females and males were placed into a hollow black cylinder that had been placed in the center of a cage and the lids of these tubes were opened and no longer closed to allow them to escape. Talc powder was uniformly daubed around the interior of the black cylinder to ensure that adults could not climb out. To provide illumination, a 30 W fluorescent light was directed at the top of the cage from a distance of 20 cm. The collection of fliers was performed every 12 h. The experiment was performed until all parasitoid wasps were dead. The flight capacity was calculated as (the total number of *P. incisi* flew out the hollow black cylinder / the total number of *P. incisi* used) × 100%. For each treatment, nine replicates of this experiment were conducted.

2.3.3. Longevity and Reproductive Parameters

Pairs of *P. incisi* adults that had emerged within 24 h were confined in individual centrifuge tubes to observe mating behavior; this ensured that all adults used in this experiment were mated. Subsequently, each pair of *P. incisi* adults were transferred into a plastic jar (diameter: 15 cm, high: 10 cm) and the top was wrapped with gauze for ventilation. Honey and water were provided: honey was daubed on the gauze and water was absorbed in a sponge inside the plastic jar. Excessive numbers of second-instar *B. dorsalis* larvae were transferred to a small oviposition plate (diameter: 3.5 cm, high: 0.5 cm) that was covered with 80 mesh net, and then a small oviposition plate was provided for *P. incisi* adults to parasitize for 24 h. The larvae were refreshed daily until the female had died. After parasitism, larvae were reared on artificial diets until pupation. Larvae and pupae that died during development were removed for further observation and dissection under a microscope to determine whether they had been parasitized. The number of both sexes of the progeny of each treatment and control were documented daily. The pre-oviposition period, oviposition period, post-oviposition period, longevity of G1, and emergence rate as well as the female proportion of G2 were recorded for each treatment and control. A total of 15 replicates were performed for each treatment.

2.4. Statistical Analysis

All data were analyzed after checking that the data were normally distributed and there was homogeneity of variances (SPSS Inc., Chicago, IL, USA). Percentage data were arcsine square-root-transformed for further statistical analysis; however, untransformed data are presented in tables. The effects of storage temperature, storage duration, pupal age interval, and their interactions on emergence and proportion of female G1 were analyzed by univariate three-way ANOVA (generalized linear model, GLM). One-way ANOVA was conducted to analyze the effects of the pupal cold storage protocol on dry weight, flight ability, longevity, and reproduction parameters of G1 *P. incisi* adults and the emergence rate and female proportion of G2. Differences between treatments and control were assessed using a one-way ANOVA with Tukey's honestly significant difference (HSD) test ($p < 0.05$) for multiple mean comparisons.

3. Results

3.1. Effects of PupalCold Storage on the Emergence Parameters of P. incisi

3.1.1. Emergence Rate

A significant difference was observed in the emergence rate of *P. incisi* among storage temperature ($F_{3,432}$ = 1286.705, $p < 0.001$), storage duration ($F_{3,432}$ = 426.532, $p < 0.001$), pupal age interval ($F_{2,432}$ = 196.2910, $p < 0.001$), storage temperature × storage duration ($F_{9,432}$ = 12.342, $p < 0.001$), storage temperature ×pupal age interval ($F_{6,432}$ = 68.522, $p < 0.001$), storage duration × pupal age interval ($F_{6,432}$ = 6.573, $p < 0.001$), and their interactions ($F_{18,432}$ = 3.881, $p < 0.001$).

Overall, the emergence rate of *P. incisi* pupae of the same age decreased as temperature decreased and storage duration increased. Regardless of pupal age interval and storage duration, *P. incisi* pupae stored at 4, 7, and 10 °C exhibited significantly lower emergence rates compared to the control group. However, no significant differences were observed for middle-age *P. incisi* pupae stored at 13 °C for 10 days ($p = 1.000$) and 15 days ($p = 0.343$) as well as late-age *P. incisi* pupae stored at 13 °C for 10 days ($p = 1.000$) and 15 days ($p = 0.779$) in comparison with the control group. Surprisingly, after being subjected to the same cold storage protocol of 4, 7, and 10 °C, middle-age *P. incisi* pupae presented the highest emergence rate compared to the prepupae and late-age pupae (Table 1).

Table 1. Emergence rate of *P. incisi* after being subjected to different pupal cold storage treatments ($n = 9$).

Storage Temperature (°C)	Storage Time (d)	Emergence Rate (%)		
		Pupal Age Interval		
		Prepupae	Middle-Age	Late-Age
4	10	7.57 ± 1.66 c	16.84 ± 1.96 b	4.55 ± 1.13 cd
	15	1.77 ± 0.89 ef	6.64 ± 1.50 c	0 f
	20	0 f	1.14 ± 0.76 f	0 f
	25	0 f	0 f	0 f
25 (control)			83.27 ± 2.36 a	
7	10	13.71 ± 2.17 c	25.22 ± 2.74 b	11.47 ± 1.28 c
	15	3.43 ± 1.2 de	11.08 ± 0.87 c	1.53 ± 0.84 de
	20	0.51 ± 0.51 e	5.44 ± 1.63 d	0 e
	25	0 e	0 e	0 e
25 (control)			83.27 ± 2.36 a	
10	10	19.03 ± 2.02 c	45.81 ± 2.50 b	39.07 ± 3.23 b
	15	15.23 ± 2.14 cd	36.19 ± 2.34 b	20.05 ± 2.04 c
	20	9.34 ± 1.56 de	16.29 ± 2.86 cd	3.36 ± 1.11e fg
	25	2.64 ± 0.84 fg	5.46 ± 1.42 ef	0 g
25 (control)			83.27 ± 2.36 a	
13	10	37.89 ± 3.16 c	81.29 ± 2.50 a	80.62 ± 1.92 a
	15	33.52 ± 4.32 cd	73.22 ± 2.34 a	75.71 ± 3.26 a
	20	20.06 ± 3.06 d	54.80 ± 2.73 b	59.33 ± 2.44 b[E]
	25	6.01± 1.48 e	38.49 ± 2.95 c[E]	46.67± 2.64 bc[E]
25 (control)			83.27 ± 2.36 a	

Note: Data are presented as mean ± SE. Different lowercase letters indicate significant differences at the 0.05 level by Tukey's test. Control means *P. incisi* pupae developed at 25 °C. [E] means that part of *P. incisi* have emerged during the cold storage protocol.

3.1.2. Female Proportion

The proportion of females was not affected by storage temperature ($F_{3,253}$ = 0.469, $p = 0.704$), storage duration ($F_{3,253}$ = 0.698, $p = 0.554$), pupal age interval ($F_{2,253}$ = 0.027, $p = 0.973$), storage temperature × storage duration ($F_{6,253}$ = 0.490, $p = 0.816$), storage temperature $\times$ pupal age interval ($F_{5, 253} = 0.093$, $p = 0.993$), storage duration $\times$ pupal age interval ($F_{6, 253} = 0.087$, $p = 0.998$), or their interactions ($F_{3, 253} = 0.069$, $p = 0.976$). There was no significant difference between the control group and all treatments (Table 2).

Table 2. Proportion of emerging *P. incisi* females after being subjected to different pupal cold storage treatments ($n = 9$).

Storage Temperature (°C)	Storage Time (d)	Female Proportion (%)		
		Pupal Age Interval		
		Prepupae	Middle-Age	Late-Age
4	10	80.21 ± 7.56 a	69.51 ± 5.06 a	-
	15	-	72.92 ± 11.86 a	-
	20	-	-	-
	25	-	-	-
25 (control)	-	-	71.53 ± 2.02 a	
7	10	71.48 ± 6.78 a	71.42 ± 4.92 a	74.07 ± 5.97 a
	15	-	64.82 ± 10.09 a	-
	20	-	80.56 ± 7.38 a	-
	25	-	-	-
25 (control)			71.53 ± 2.02 a	
10	10	74.23 ± 6.01 a	68.21 ± 2.96 a	69.81 ± 2.85 a
	15	71.09 ± 6.17 a	71.65 ± 3.06 a	72.67 ± 4.30 a
	20	71.88 ± 8.78 a	70.95 ± 4.15 a	-
	25	-	80.95 ± 8.13 a	-
25 (control)			71.53 ± 2.02 a	
13	10	70.65 ± 5.34 a	69.20 ± 3.08 a	66.79 ± 1.99 a
	15	69.58 ± 4.01 a	70.12 ± 2.62 a	68.43 ± 1.98 a
	20	73.70 ± 5.78 a	69.23 ± 4.06 a	71.56 ± 3.21 a^E
	25	68.75 ± 15.27 a	70.56 ± 4.26 a^E	69.42 ± 2.43 a^E
25 (control)			71.53 ± 2.02 a	

Note: Data are presented as mean $\pm$ SE. Different lowercase letters indicate significant differences at the 0.05 level by Tukey's test. Control means *P. incisi* pupae developed at 25 °C. ^{E} means that part of *P. incisi* have emerged during the cold storage protocol. "-" means that the emergence rate was less than 5% after pupal cold storage, and therefore excluded from the analysis.

3.2. Effects of Pupal Cold Storage on the Quality of P. incisi Adults

3.2.1. Dry Weight

Pupal cold storage lead to significant impacts on the dry weight of both sexes of post-storage *P. incisi* adults according to one-way ANOVA (female: $F_{4, 40} = 36.795$, $p < 0.001$; male: $F_{4, 40} = 22.323$, $p < 0.001$). Furthermore, based on the result of Tukey's HSD test, except for the cold storage 4 (CS4) protocol (female: $p < 0.05$; male: $p < 0.05$), all other pupal cold storage treatments did not differ from the control for the dry weight of both sex adults (Table 3).

3.2.2. Flight Capacity

The flight capacity of post-storage *P. incisi* adults was significantly affected by the cold storage protocol (female: $F_{4, 40} = 3.520$, $p < 0.05$; male: $F_{4, 40} = 2.926$, $p < 0.05$). CS4 lead to the lowest flight capacity of both sex adults and was significantly inferior to the control (female: $p < 0.05$; male: $p < 0.05$) (Table 3).

Table 3. The dry weight of G1 *P. incisi* post-storage females (a) and males (b) that had been subjected to different pupal cold storage treatments (*n* = 9).

Treatment	Dry Weight (mg)		Flight Ability (%)	
	Female	**Male**	**Female**	**Male**
Control (25 °C)	59.67 ± 0.76 a	40.30 ± 0.59 a	69.33 ± 3.65 a	62.67 ± 2.75 a
CS1	59.05 ± 0.88 ab	40.55 ± 0.38 a	67.56 ± 4.15 ab	60.22 ± 2.99 ab
CS2	59.15 ± 0.59 ab	40.10 ± 0.50 ab	60.22 ± 1.61 ab	54.44 ± 2.97 ab
CS3	57.72 ± 0.60 ab	39.24 ± 0.34 ab	61.33 ± 1.91 ab	55.11 ± 2.21 ab
CS4	56.46 ± 0.97 b	38.19 ± 0.64 b	56.67 ± 1.86 b	52.00 ± 1.70 b

Note: CS1: Cold storage 1, late-age pupae stored at 13 °C for 10 d; CS2: cold storage 2, middle-age pupae stored at 13 °C for 10 d; CS3: cold storage 3, late-age late pupae stored at 13 °C for 15 d; CS4: cold storage 4, middle-age pupae stored at 13 °C for 15 d. Control means *P. incisi* pupae developed at 25 °C. Bars topped by the same letter do not differ significantly (*p* > 0.05) according to Tukey's HSD test (one-way ANOVA).

3.2.3. Longevity

For female parasitoids, pupal cold storage had significant effects on longevity (female: $F_{4, 70} = 3.795$, $p < 0.01$), and a remarkable reduction in female longevity was observed for CS4 ($p < 0.05$). However, the longevity of males was not significantly influenced by different treatments ($F_{4, 70} = 1.669$, $p = 0.167$) (Figures 1 and 2).

3.2.4. G1 Reproductive Performance and G2 Emergence Parameters

Pre-oviposition period ($F_{4, 70} = 0.353$, $p = 0.841$), post-oviposition period ($F_{4, 70} = 2.128$, $p = 0.086$), G2 emergence rate ($F_{4, 70} = 0.412$, $p = 0.799$), and G2 female proportion ($F_{4, 70} = 1.864$, $p = 0.127$) were not significantly affected by pupal cold storage. However, significant effects were observed for total offspring per female ($F_{4, 70} = 15.245$, $p < 0.001$), daily offspring ($F_{4, 70} = 3.625$, $p < 0.05$), and oviposition period ($F_{4, 70} = 3.893$, $p < 0.001$). The highest total offspring produced by a female was in the control group, and there was a significant difference between the control in comparison to females subjected to CS2 ($p < 0.05$) and CS4 ($p < 0.01$). Furthermore, CS4 resulted in significantly lower daily offspring ($p < 0.05$) and a shorter oviposition period ($p < 0.05$) than the control (Tables 4 and 5).

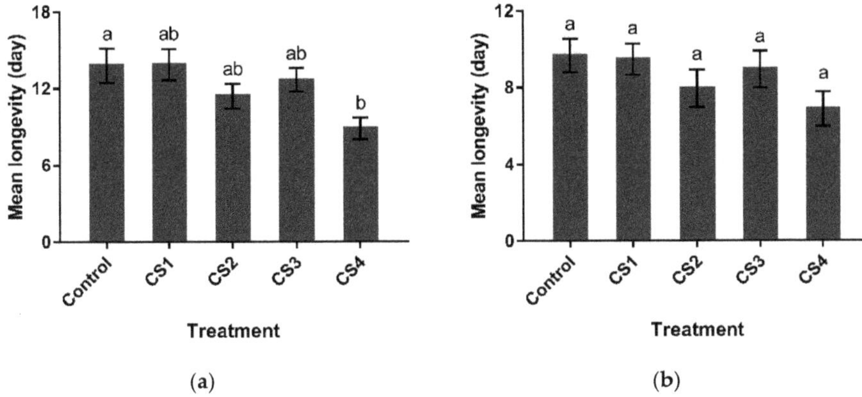

(a)   (b)

Figure 1. Longevity of G1 *P. incisi* post-storage females (**a**) and males (**b**) that had been subjected to different pupal cold storage treatments. CS1: cold storage 1, late-age pupae stored at 13 °C for 10 d; CS2: cold storage 2, middle-age pupae stored at 13 °C for 10 d; CS3: cold storage 3, late-age late pupae stored at 13 °C for 15 d; CS4: cold storage 4, middle-age pupae stored at 13 °C for 15 d. Control means *P. incisi* pupae developed at 25 °C. Bars topped with the same letter do not differ significantly (*p* > 0.05) according to Tukey's HSD test (one-way ANOVA) (*n* = 15).

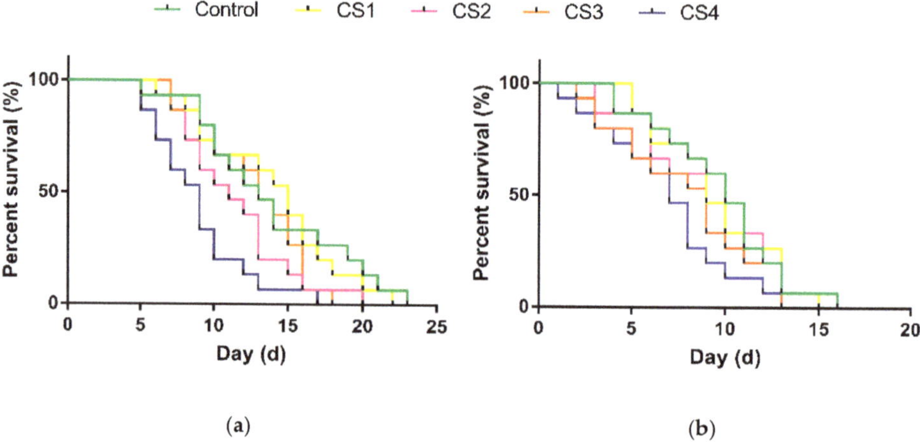

(a) (b)

Figure 2. Survival curves of *P. incisi* post-storage female (**a**) and male (**b**) that had been subjected to different pupal cold storage treatments. CS1: cold storage 1, late-age pupae stored at 13 °C for 10 d; CS2: cold storage 2, middle-age pupae stored at 13 °C for 10 d; CS3: cold storage 3, late-age late pupae stored at 13 °C for 15 d; CS4: cold storage 4, middle-age pupae stored at 13 °C for 15 d. Control means *P. incisi* pupae developed at 25 °C (*n* = 15).

Table 4. Reproductive parameters of G1 *P. incisi* post-storage adults (*n* = 15).

Treatment	Pre-Oviposition Period (d)	Oviposition Period (d)	Post-Oviposition Period (d)	Total No. of Offspring	No. of Daily Offspring
25 °C (control)	0.33 ± 0.13 a	10.73 ± 0.73 a	2.73 ± 0.85 a	69.13 ± 4.44 a	6.72 ± 0.41 a
CS1	0.27 ± 0.15 a	11.07 ± 0.72 a	2.53 ± 0.56 a	68.27 ± 3.75 ab	6.28 ± 0.45 ab
CS2	0.20 ± 0.11 a	9.60 ± 0.66 ab	1.60 ± 0.42 a	54.80 ± 5.51 b	5.53 ± 0.37 ab
CS3	0.27 ± 0.12 a	10.40 ± 0.58 ab	2.00 ± 0.48 a	65.60 ± 3.84 ab	6.46 ± 0.31 a
CS4	0.40 ± 0.13 a	7.73 ± 0.67 b	0.73 ± 0.25 a	37.07 ± 4.59 c	4.92 ± 0.40 b

Note: Data are presented as mean ± SE. Different lowercase letters indicate significant differences at the 0.05 level by Tukey's test. CS1: cold storage 1, late-age pupae stored at 13 °C for 10 d; CS2: cold storage 2, middle-age pupae stored at 13 °C for 10 d; CS3: cold storage 3, late-age late pupae stored at 13 °C for 15 d; CS4: cold storage 4, middle-age pupae stored at 13 °C for 15 d. Control means *P. incisi* pupae developed at 25 °C.

Table 5. Emergence rate and female proportion of progeny (G2) (*n* = 15).

Treatment	G2 Emergence Rate (%)	G2 Female Proportion (%)
Control (25 °C)	85.51 ± 1.31 a	71.36 ± 1.44 a
CS1	83.75 ± 1.20 a	69.43 ± 1.13 a
CS2	83.45 ± 1.52 a	70.73 ± 0.95 a
CS3	83.96 ± 1.11 a	71.02 ± 0.73 a
CS4	84.23 ± 1.21 a	67.26 ± 1.58 a

Note: Data are presented as mean ± SE. Different lowercase letters indicate significant differences at the 0.05 level by Tukey's test. CS1: cold storage 1, late-age pupae stored at 13 °C for 10 d; CS2: cold storage 2, middle-age pupae stored at 13 °C for 10 d; CS3: cold storage 3, late-age late pupae stored at 13 °C for 15 d; CS4: cold storage 4, middle-age pupae stored at 13 °C for 15 d. Control means *P. incisi* pupae developed at 25 °C.

4. Discussion

The utilization of opiine parasitoids, especially *F. arisanu*, has successfully alleviated the serious damage of *B. dorsalis* to fruit products in Hawaii, which inspired other regions to adopt a biological control program for managing this pest [21]. However, today, *F. arisanu* has only been recorded in Zhanjiang City, Guangdong Province in China [22], which means that *F. arisanu* may not be able to be the dominant bioagent against *B. dorsalis* in some regions of China. Furthermore, considering the various ecological conditions in different regions of mainland China, it is indispensable to dig local parasitoid resources to develop the most suitable local biological control programs against *B. dorsalis*. Previous research in Zhangzhou City, Fujian Province, China, indicated that there were four parasitic wasp species against *B. dorsalis*, including *P. incisi*, *Pachycrepoideus vindemmiae* (Rondani), *Pachycrepoideus vindemmiae* (Rondani), and *Spalangia endius* (Walker) (Hymenoptera: Pteromalidae), of which *P. incisi* occupied a dominant proportion (77.6%) of this parasitic wasp population [9]. Therefore, *P. incisi* is a highly suitable bioagent for biological control programs against *B. dorsalis* in that region of China.

The present study is the first, to our knowledge, that has aimed to optimize the cold storage protocol of *P. incisi* to improve the efficiency and flexibility of biological control programs against *B. dorsalis*. Our results demonstrate that storage temperature, storage duration, pupal age interval, storage temperature × storage duration, storage temperature ×pupal age interval, and storage duration × pupal age interval and their interactions have significant impacts on the viability of immature *P. incisi*. However, when late-age *P. incisi* pupae were subjected to CS1 and CS3, the emergence parameters of both G1 and G2 progeny, and G1 quality parameters (including flight ability, dry weight, longevity, and reproduction parameters) did not differ significantly from the control treatment.

The essence of cold storage is to utilize a sub-optimum temperature to prolong the developmental time of a bioagent whilst maintaining its quality and effectiveness against a pest [23]. Cold storage protocols are extremely valuable for inundative biological control, given that they require large-scale production and the release of huge numbers of bioagents. However, modulating the development of bioagents via cold storage can pose lethal and sub-lethal effects to their survivorship [24,25]. In the present study, a reduction in the storage temperature or an extension of the storage period resulted in a decrease in the emergence rate of *P. incisi*. This is consistent with previous research on the cold storage of two other braconid parasitoids that are used against *B. dorsalis*, namely *F. arisanus* and *D. longicaudata*; these studies found that the viability of parasitized pupae was gradually reduced as the temperature decreased or the storage period was extended [16,17]. In addition, we interestingly found that middle-age *P. incisi* pupae exhibited superior performance in the emergence rate in comparison to prepupae and late-age pupae subjected to the same cold storage protocols of 4, 7, or 10 °C. Similarly, a study on *Encarsia formosa* (Gahan) (Hymenoptera: Aphelinidae) indicated that both early- and late-stage pupae exhibited inferior tolerance to low temperatures in comparison with mid-stage pupae, with the mid-stage pupae exhibiting higher survival rates [26].

Although cold storage clearly has advantages in enhancing the efficiency and flexibility of mass-rearing and release programs, the undesirable effects on the quality of post-storage insects are of concern. A significant amount of research has emphasized that subjecting bioagents to cold conditions can result in cold stress and excessive consumption of energy reserves, thereby influencing their quality parameters and their successful application under field conditions [11,27]. Our results demonstrated that the cold storage of *P. incisi* pupae was deleterious to the dry weight of post-storage adults. Lins et al. reported that the body mass loss was directly proportional to the extension in the storage period when prepupae of *Praon volucre* (Haliday) (Hymenoptera: Braconidae) were stored at 5 °C [28]. These weight losses during cold storage may chiefly result from the immature parasitoid wasps consuming a body of energy and lipid reserves to ensure survival during the period of low temperature. Excessive lipid loss is fatal to parasitoid adults as they cannot synthesize lipids by themselves, and thereby perform a trade-off between survivorship and fecundity, in turn affecting their lifespan [29–31]. In our study, the longevity of female *P. incisi* that emerged from middle-age pupae stored at 13 °C for 15 days was significantly shorter. This is in accordance with results found in *D. longicaudata* [17], which revealed that females that emerged from parasitized *B. doralis* pupae stored below 8 °C had a shorter lifespan than the control, regardless of the storage period. Likewise, the cold storage of other braconid wasps, such as *Aphidius ervi* (Haliday), *Aphidius picipes* (Nees), and *Bracon hebetor* (Say) (Hymenoptera: Braconidae), during the pupal stage, indicated a similar tendency in the longevity of post-storage adults [32–34].

Dispersibility, one of the most important quality parameters for parasitoids when considering the various, complex, and harsh field conditions they need to withstand whilst host-seeking, hiding from predators, and searching for resting places, is particularly vulnerable to the effects of low-temperature storage. In fact, the neuro-muscular dysfunction that can be induced by cold storage directly affects the bioagent's ability to disperse, and is a major obstacle to the practical application of post-storage insects in the field [10,35]. In concordance with this, our results demonstrated that subjecting middle-age *P. incisi* pupae to 15 days of cold storage at 13 °C resulted in the emerged adults displaying inferior flight ability. A study on *E. formosa* and *Encarsia eremicus* (Rose) (Hymenoptera: Aphelinidae) indicated that increasing the cold exposure period of pupae resulted in a reduction in the flight capacity of the emerged adults [26]. Similarly, *P. volucre* pre-pupae developed at a sub-ambient temperature of 5 °C exhibited lower flight capacity than the control group [28].

In addition to dispersibility, the reproductive system of bioagents is extremely susceptible to sub-optimum conditions. In our study, for middle-age *P. insici* pupae subjected to cold storage at 13 °C for 10 or 15 days, the number of progeny produced by post-storage adults was remarkably decreased. This is consistent with previous research on braconid parasitoids, whereby a reduction in fecundity was strongly associated with low temperature and storage period [28,33,36]. Furthermore, the reproductive parameters of *P. incisi* that originated from the Zhangzhou region showed a significant difference to the Hawaii *P. incisi* strain [37]. Liang et al. [38] indicated that there was a certain difference between two geographic *P. incisi* populations by using random amplified polymorphic DNA analysis, this may account for the remarkable difference in reproductive parameters between these two strains.

The viability of mass-release biocontrol programs is largely determined by the ability to generate sufficient reserves of the bioagent in a relatively short timeframe. Furthermore, flexibility is also required in both the rearing and release schedules to deal with unforeseen factors such as adverse weather and transportation delays. As such, the cold storage technique can help mitigate these problems [39] and is increasingly undertaken as a support approach in classical biological control programs [11]. Nonetheless, the adverse effects induced by sub-ambient temperature remarkably reduce the performance of post-storage bioagents. This consequently leads to challenges in balancing the logistics of the release program with field performance. In order to minimize the losses in quantity and fitness, recent studies have primarily concentrated on optimizing the cold storage protocol of the bioagent. As such, our present study aims to provide a foundation for optimizing the cold storage technique of *P. incisi* and our results demonstrate that *P. incisi* pupae subjected to CS1 and CS3 does not result in significant adverse effects on the emergence rate and quality of post-storage adults. Such information is vital for the mass production and release of *P. incisi* as a dominant biological control agent against *B. dorsalis* in that region of China. In addition, in this study, we used microscopic examination to distinguish parasitized puparia and unparasitized puparia to facilitate scientific research. However, for the mass storage of *P. incisi* pupae for release, we still propose using physical approaches, such as utilizing the suitable mesh of net that prevents flies from escaping, while without restricting the emerged *P. incisi*. Further studies will be carried out to assess the potential control efficacy of post-storage *P. insici* against the *B. dorsalis* population under field conditions.

Author Contributions: Methodology, J.L., C.L., and Q.J.; performed the experiments, J.L., D.Y., X.H., Y.G., and S.S.; writing—original draft preparation, J.L. and P.C.; writing—review and editing, J.L. and P.C., supervision, C.L. and Q.J. All authors have read and agreed to the published version of the manuscript.

Funding: This work was funded by the Industry University Research Project of Fujian Science and Technology Department (2019N5003) and the National Key R&D Program of China (2017YFD0202000).

Institutional Review Board Statement: Not applicable.

Informed Consent Statement: Not applicable.

Data Availability Statement: The data presented in this study are available from the corresponding author on reasonable request.

Acknowledgments: We extend our sincerest appreciation to Shumei Wang for giving us critical suggestions regarding this research.

Conflicts of Interest: The authors declare no conflict of interest. The funders had no role in the design of the study; in the collection, analyses, or interpretation of data; in the writing of the manuscript, or in the decision to publish the results.

References

1. Liu, H.; Zhang, D.J.; Xu, Y.J.; Wang, L.; Cheng, D.F.; Qi, Y.X.; Zeng, L.; Lu, Y.Y. Invasion, expansion, and control of *Bactrocera dorsalis* (Hendel) in China. *J. Integr. Agric.* **2019**, *18*, 771–787. [CrossRef]
2. Clarke, A.R.; Li, Z.H.; Qin, Y.J.; Zhao, Z.H.; Liu, L.J.; Schutze, M.K. *Bactrocera dorsalis* (Hendel) (Diptera: *Tephritidae*) is not invasive through Asia: It's been there all along. *J. Appl. Entomol.* **2019**, *143*, 797–801. [CrossRef]
3. Mutamiswa, R.; Nyamukonduwa, C.; Chikowore, G.; Chidawanyika, F. Overview of oriental fruit fly, *Bactrocera dorsalis* (Hendel) (Diptera: *Tephritidae*) in Africa: From invasion, bio-ecology to sustainable management. *Crop. Prot.* **2021**, *141*, 105402. [CrossRef]
4. Piñero, J.C.; Souder, S.K.; Smith, T.R.; Vargas, R.I. Attraction of *Bactrocera cucurbitae* and *Bactrocera dorsalis* (Diptera: *Tephritidae*) to beer waste and other protein sources laced with ammonium acetate. *Fla. Entomol.* **2017**, *100*, 70–76. [CrossRef]
5. Jin, T.; Zeng, L.; Lin, Y.Y.; Lu, Y.Y.; Liang, G.W. Insecticide resistance of the oriental fruit fly, *Bactrocera dorsalis* (Hendel) (Diptera: *Tephritidae*), in mainland China. *Pest. Manag. Sci.* **2011**, *67*, 370–376. [CrossRef]
6. Diaz-Fleischer, F.; Perez-Staples, D.; Cabrera-Mireles, H.; Montoya, P.; Liedo, P. Novel insecticides and bait stations for the control of *Anastrepha* fruit flies in mango orchards. *J. Pestic. Sci.* **2017**, *90*, 1–8. [CrossRef]
7. Yang, J.Q.; Cai, P.M.; Chen, J.; Zhang, H.H.; Wang, C.; Xiang, H.J.; Yang, Y.C.; Chen, J.H.; Ji, Q.E.; Song, D.B. Interspecific competition between Fopius arisanus and Psyttalia incisi (Hymenoptera: *Braconidae*), parasitoids of Bactrocera dorsalis (Diptera: *Tephritidae*). *Biol. Control* **2018**, *40*, 183–189. [CrossRef]
8. Liang, G.H.; Chen, J.H.; Huang, J.C. The functional response and disturbance response of larvae of *Psyttalia incisi* (Silvestri) to oriental fruit flies. *Acta Agric. Univ. Jiangxiensis* **2006**, *28*, 200–203.
9. Liang, G.H.; Wu, Y.; Chen, J.H. Seasonal incidence of *Bactrocera dorsalis* and its parasitoids in field. *J. Southwest For. Coll.* **2006**, *26*, 72–74.
10. Colinet, H.; Boivin, G. Insect parasitoids cold storage: A comprehensive review of factors of variability and consequences. *Biol. Control* **2011**, *58*, 83–95. [CrossRef]
11. Rathee, M.; Ram, P. Impact of cold storage on the performance of entomophagous insects: An overview. *Phytoparasitica* **2018**, *46*, 421–449. [CrossRef]
12. Rezaei, M.; Talebi, A.A.; Fathipour, Y.; Karimzadeh, J.; Mehrabadi, M.; Reddy, G.V.P. Effects of cold storage on life-history traits of *Aphidius matricariae*. *Entomol. Exp. Appl.* **2020**, *168*, 800–807. [CrossRef]
13. Ghazy, N.A.; Suzuki, T.; Amano, H.; Ohyama, K. Air temperature optimisation for humidity-controlled cold storage of the predatory mites *Neoseiulus californicus* and *Phytoseiulus persimilis* (Acari: *Phytoseiidae*). *Pest. Manag. Sci.* **2014**, *70*, 483–487. [CrossRef] [PubMed]
14. Lü, X.; Han, S.C.; Li, J.; Liu, J.S.; Li, Z.G. Effects of cold storage on the quality of *Trichogramma dendrolimi* Matsumura (*Hymenoptera: Trichogrammatidae*) reared on artificial medium. *Pest. Manag. Sci.* **2019**, *75*, 1328–1338. [CrossRef] [PubMed]
15. Daane, K.M.; Wang, X.G.; Johnson, M.W.; Cooper, M.L. Low temperature storage effects on two olive fruit fly parasitoids. *BioControl* **2013**, *58*, 175–185. [CrossRef]
16. Long, X.Z.; Chen, K.W.; Xian, J.D.; Lu, Y.Y.; Zeng, L. Cold storage technique of *Diachasmimorpha longicaudata* (Ashmead). *J. Environ. Entomol.* **2014**, *36*, 115–121.
17. Wang, H.L. Effects of Spinosad and Temperature and Humidity on the Parasitoid *Fopius arisanus* (Sonan). Master's Thesis, Fujian Agricultural and Forestry University, Fuzhou, China, 2011.
18. Ji, Q.E.; Dong, C.Z.; Chen, J.H. A new record species—*Opius incisi* Silvestri (*Hymenoptera: Braconidae*) parasitizing on *Dacus dorsalis* (Hendal) in China. *Entomotaxonomia* **2004**, *26*, 144–145.
19. Chang, C.L.; Vargas, R.I.; Caceres, C.; Jang, E.; Cho, I.K. Development and assessment of a liquid larval diet for *Bactrocera dorsalis* (Diptera: *Tephritidae*). *Ann. Entomol. Soc. Am.* **2006**, *99*, 1191–1198. [CrossRef]
20. Wang, X.G.; Messing, R.H. Potential interactions between pupal and egg-or larval-pupal parasitoids of Tephritid fruit flies. *Environ. Entomol.* **2004**, *33*, 1313–1320. [CrossRef]
21. Vargas, R.I.; Leblanc, L.; Harris, E.; Manoukis, N.C. Regional suppression of *Bactrocera* fruit flies (Diptera: *Tephritidae*) in the Pacific through biological control and prospects for future introductions into other areas of the world. *Insects* **2012**, *3*, 729–742. [CrossRef] [PubMed]
22. Yao, M.J.; Xie, C.H.; He, Y.B.; Qiu, B.; Chen, H.Y.; Xu, Z.F. Investigation on hylmenopterous parasitoids of *Bactrocera dorsalis* (Hendel) in Guangdong. *J. Environ. Entomol.* **2008**, *30*, 350–356.
23. Benelli, M.; Ponton, F.; Lallu, U.; Mitchell, K.A.; Taylor, P.W. Cool storage of Queensland fruit fly pupae for improved management of mass production schedules. *Pest. Manag. Sci.* **2019**, *75*, 3184–3192. [CrossRef]
24. Sakaki, S.; Jalali, M.A.; Kamali, H.; Nedved, O. Effect of low-temperature storage on the life history parameters and voracity of *Hippodamia variegata* (Coleoptera: *Coccinellidae*). *Eur. J. Entomol.* **2019**, *116*, 10–15. [CrossRef]

25. Benelli, M.; Ponton, F.; Taylor, P.W. Cool sotrage of Queensland fruit fly eggs for increased flexibility in rearing programs. *Pest. Manag. Sci.* **2019**, *75*, 1056–1064. [CrossRef]

26. Luczynski, A.; Nyrop, J.P.; Shi, A. Influence of cold storage on pupal development and mortality during storage and on post-storage performance of *Encarsia formosa* and *Eretmocerus eremicus* (Hymenoptera: Aphelinidae). *Biol. Control* **2006**, *40*, 107–117. [CrossRef]

27. Neven, L.G.; Hansen, L.D. Effects of temperature and controlled atmospheres on codling moth metabolism. *Ann. Entomol. Soc. Am.* **2010**, *103*, 418–423. [CrossRef]

28. Lins, J.C.; Bueno, V.H.P.; Sidney, L.A.; Silva, D.B.; Sampaio, M.V.; Pereira, J.M.; Nomelini, Q.S.S.; van Lenteren, J.C. Cold storage affects mortality, body mass, lifespan, reproduction and flight capacity of *Praon volucre* (*Hymenoptera: Braconidae*). *Eur. J. Entomol.* **2013**, *110*, 263–270. [CrossRef]

29. Jervis, M.A.; Ferns, P.N.; Heimpel, G.E. Body size and the timing of egg production in parasitoid wasps: A comparative analysis. *Funct. Ecol.* **2003**, *17*, 375–383. [CrossRef]

30. Colinet, H.; Boivin, G.; Hance, T. Manipulation of parasitoid size using the temperature-size rule: Fitness consequences. *Oecologia* **2007**, *152*, 425–433. [CrossRef]

31. Visser, B.; le Lann, C.; den Blanken, F.J.; Harvey, J.A.; van Alphen, J.M.; Ellers, J. Loss of lipid synthesis as an evolutionary consequence of a parasitic lifestyle. *Proc. Natl. Acad. Sci. USA* **2010**, *107*, 8677–8682. [CrossRef] [PubMed]

32. Ismail, M.; van Baaren, J.; Hance, T.; Perre, J.S.; Vernon, P. Stress intensity and fitness in the parasitoid *Aphidius ervi* (*Hymenoptera: Braconidae*): Temperature below the development threshold combined with a fluctuating thermal regime is a must. *Ecol. Entomol.* **2013**, *38*, 355–363. [CrossRef]

33. Alam, M.S.; Alam, M.Z.; Alam, S.N.; Miah, M.R.U.; Mian, M.I.H. Effect of storage duration on the stored pupae of parasitoid *Bracon hebetor* (Say) and its impact on parasitoid quality. *Bangladesh J. Agric. Res.* **2016**, *41*, 297–310. [CrossRef]

34. Amice, G.; Vernon, P.; Outreman, Y.; van Alphen, J.; van Baaren, J. Variability in responses to thermal stress in parasitoids. *Ecol. Entomol.* **2008**, *33*, 701–708. [CrossRef]

35. Yocum, G.D.; Zdárek, J.; Joplin, K.H.; Lee, R.E.; Smith, D.C.; Manter, K.D.; Denlinger, D.L. Alteration of the eclosion rhythm and eclosion behavior in the flesh fly, Sarcophaga crassipalpis, by low and high temperature stress. *J. Insect. Physiol.* **1994**, *40*, 13–21. [CrossRef]

36. Yan, Z.; Yue, J.J.; Bai, C.; Peng, Z.Q.; Zhang, C.H. Effects of cold storage on the biological characteristics of *Microplitis prodeniae* (Hymenoptera: Braconidae). *Bull. Entomol. Res.* **2017**, *107*, 506–512. [CrossRef] [PubMed]

37. Vargas, R.I.; Ramadan, M.; Hussain, T.; Mochizuki, N.; Bautista, R.C.; Stark, J.D. Comparative demography of six fruit fly (Diptera: Tephritidae) parasitoids (Hymenoptera: Braconidae). *Biol. Control* **2002**, *25*, 30–40. [CrossRef]

38. Liang, G.H.; Huang, J.C.; Chen, J.H. Comparative analysis on RAPD of two geographic populations of *Psyttalia incisi. J. Fujian Coll. For.* **2007**, *27*, 16–19.

39. Leopold, R. Colony maintenance and mass-rearing: Using cold storage technology for extending the shelf-life of insects. In *Area-Wide Control of Insect Pests: From Research to Field Implementation*, 2nd ed.; Vreysen, M.J.B., Robinson, A.S., Hendrichs, J., Eds.; Springer: Dordrecht, The Netherlands, 2007; pp. 149–162.

Article

Gene Characterization and Enzymatic Activities Related to Trehalose Metabolism of In Vitro Reared *Trichogramma dendrolimi* Matsumura (Hymenoptera: Trichogrammatidae) under Sustained Cold Stress

Xin Lü *, Shi-chou Han, Zhi-gang Li, Li-ying Li and Jun Li *

Guangdong Key Laboratory of Animal Conservation and Resource Utilization,
Guangdong Public Laboratory of Wild Animal Conservation and Utilization, Institute of Zoology,
Guangdong Academy of Sciences, 105 Xingang Road West, Guangzhou 510260, China;
hansc@giabr.gd.cn (S.-c.H.); leegdei@163.com (Z.-g.L.); liyingl32@163.com (L.-y.L.)
* Correspondence: greenhopelv@163.com (X.L.); junl@giabr.gd.cn (J.L.)

Received: 23 September 2020; Accepted: 4 November 2020; Published: 7 November 2020

Simple Summary: Trehalose is a non-reducing disaccharide that presents in a wide variety of organisms, where it serves as an energy source or stress protectant. Trehalose is the most characteristic sugar of insect hemolymph and plays a crucial role in the regulation of insect growth and development. *Trichogramma* species are economically important egg parasitoids, which are being mass-produced for biological control programs worldwide. Many *Trichogramma* species could be mass reared on artificial mediums (not insect eggs), in which components contain insect hemolymph and trehalose. These in vitro-reared parasitoid wasps were strongly affected by cold storage, but prepupae could be successfully stored at 13 °C for up to 4 weeks. The aims of the present study were to determine the role of trehalose and the relationship between trehalose and egg parasitoid stress resistance. Our study revealed that (1) trehalose regulated the growth under sustained cold stress; (2) prepupal stage is a critical developmental period and 13 °C is the cold tolerance threshold temperature; (3) in vitro reared *Trichogramma dendrolimi* could be reared at temperatures of 16 °C, 20 °C, and 23 °C to reduce rearing costs. This finding identifies a low cost, prolonged development rearing method for *T. dendrolimi*, which will facilitate improved mass rearing methods for biocontrol.

Abstract: *Trichogramma* spp. is an important egg parasitoid wasp for biocontrol of agriculture and forestry insect pests. Trehalose serves as an energy source or stress protectant for insects. To study the potential role of trehalose in cold resistance on an egg parasitoid, cDNA for trehalose-6-phosphate synthase (TPS) and soluble trehalase (TRE) from *Trichogramma dendrolimi* were cloned and characterized. Gene expressions and enzyme activities of *TdTPS* and *TdTRE* were determined in larvae, prepupae, pupae, and adults at sustained low temperatures, 13 °C and 16 °C. *TdTPS* and *TdTRE* expressions had similar patterns with higher levels in prepupae at 13 °C and 16 °C. *TdTPS* enzyme activities increased with a decrease of temperature, and *TdTRE* activity in prepupae decreased sharply at these two low temperatures. In vitro reared *T. dendrolimi* could complete entire development above 13 °C, and the development period was prolonged without cold injury. Results indicated trehalose might regulate growth and the metabolic process of cold tolerance. Moreover, 13 °C is the cold tolerance threshold temperature and the prepupal stage is a critical developmental period for in vitro reared *T. dendrolimi*. These findings identify a low cost, prolonged development rearing method, and the cold tolerance for *T. dendrolimi*, which will facilitate improved mass rearing methods for biocontrol.

Keywords: trehalase; trehalose metabolism; in vitro rearing; cold stress; *Trichogramma*

---

1. Introduction

Sugars are used for energy production and are stored as glycogen in the body fat, or as trehalose in the hemolymph [1–3]. *Trichogramma* are egg parasitoid wasps that obtain diverse nutrients, including sugars, from their host eggs during development. *Trichogramma dendrolimi* Matsumura is an important biological control agent that has been mass produced on eggs of *Corcyra cephalonica* (Stainton) and *Antheraea pernyi* (Guérin-Méneville) for biological control programs in China [4,5]. Artificial host eggs are now used to mass produce *T. dendrolimi* [6,7]. Lü et al. [8–10] developed an artificial medium containing trehalose for the continuous rearing of *T. dendrolimi* and revealed that trehalose was an essential ingredient of the artificial media. Biochemical characteristics, including trehalose content and trehalase activity in *T. dendrolimi*, continuously reared on artificial medium (in vitro) versus those reared on *A. pernyi* eggs (in vivo), were also studied [11]. The quality of in vitro reared *T. dendrolimi* was strongly affected by cold storage, but prepupae could be successfully stored at 13 °C for up to 4 weeks [12].

The developmental temperature threshold can vary, not only among insects, but also among populations [13,14]. *Trichogramma* spp. show different reactions to low temperatures when reared on different hosts or media [12,13]. For *T. dendrolimi*, the developmental threshold temperature was different among geographical populations and hosts: 10.34 °C for south population reared on *Philosamia cynthia ricini* [13]; 10.1 °C for the south population reared on *C. cephalonica* [14]; 5.34 °C/5.1 °C, 5.82 °C/5.42 °C, 11.03 °C/14.83 °C, and 12.37 °C/11.58 °C for the north population (egg, larva, prepupa and pupa) reared on *A. pernyi* [15,16]. The north population reared on *A. pernyi* was unable to complete the entire development (stopped developing at prepupal stage) at 10 °C, but was able to complete the entire development above 15 °C [15]. Compared with the in vivo (on *A. pernyi*) reared *Trichogramma*, in vitro (on artificial medium) reared *Trichogramma* of the south population was more affected by cold storage [12]. All of these experimental populations have been lab adapted. Limited information is available about inducing cold stress on the lab adapted strain of this particular insect species, and molecular mechanisms of trehalose metabolism, or the relationship between trehalose and egg parasitoid stress resistance [17–21]. Based on the facile sampling and reproducibility of the condition of in vitro reared *T. dendrolimi*, trehalose metabolism related enzymes were explored by investigating their changes in gene expression and corresponding enzyme activities to study the effect of cold stress conditions on egg parasitoids.

Trehalose is involved in the regulation of parasitoid growth and development. The disaccharide sugar trehalose serves as an energy source or stress protectant for parasitoids [22,23]. It promotes longevity, fecundity [24,25], and cold tolerance [18,26,27], and provides' energy needed to search for, and parasitize, hosts [28]. Trehalose is synthesized by trehalose-6-phosphate synthase (TPS, EC 2.4.1.15) and trehalose-6-phosphate phosphatase (TPP, EC 3.1.3.12) in the body fat, and is hydrolyzed by trehalase (TRE, EC 3.2.1.28) to yield two glucose molecules in the hemolymph [19,20,29–31]. The activity of these three enzymes also affects insect physiology and development. Trehalose and trehalase are closely associated with growth and development throughout insect life cycles [2,23]. However, we should notice the adult yields and quality of in vitro rearing some parasitoids (e.g., the tachinid *Exorista larvarum*) did not drop when the artificial medium without insect material, which trehalose has been replaced with sucrose and sucrose, was even deleted, without drops in adult yields [32,33].

TPS and TPP genes have two functional conserved domains similar to yeast genes and are homologs of yeast Tps1 (Ots A) and Tps2 (Ots B), respectively [34]. In insects, TPS is a fused gene [35] and two exons are involved in encoding trehalose synthetase [36]. Trehalase catalyses, the irreversible hydrolysis of trehalose to glucose, which is the only known pathway of trehalose utilization [2]. TRE is essential for energy metabolism and is important in insect growth and molting [37]. Trehalose may function as a cryoprotectant to stabilize proteins at low temperatures [2,18], and may also protect insects from external interference, assist in successful completion of metamorphosis, and aid survival in adverse environments [38]. In addition, trehalose is important in the regulation of insect growth and development and serves as an energy source and stress protectant [19,38–41].

TPS has been cloned, characterized, and purified from many insect species. It was first cloned from *Drosophila melanogaster* [42]. TRE has been identified in insect species, such as *Rhodnius prolixus, Nilaparvata lugens, Spodoptera exigua, Omphisa fuscidentalis,* and *Harmonia axyridis* [19,30,43–46]. Although TPS and TRE are important key enzymes for many insects, in parasitoid wasps, only the trehalase cDNA from *Pimpla hypochondriaca* has been cloned [47]. No trehalose metabolism genes in *Trichogramma*, or even in any egg parasitoids, have been characterized. To understand the role of trehalose in the cold resistance of in vitro reared *T. dendrolimi*, two genes (*TdTPS* and *TdTRE*) were identified, and cloned the full-length cDNA of *T. dendrolimi* reared on artificial medium using transcriptome data from *T. dendrolimi* reared on eggs of *A. pernyi*. The changes in gene expression and corresponding enzyme activities in four developmental stages (larva, prepupa, pupa, and adult) as each stage developed at low temperatures were also recorded. The trehalose metabolism in the life cycle of *T. dendrolimi* was systematically investigated in relation to cold hardiness.

2. Materials and Methods

2.1. Insects

Trichogramma dendrolimi were provided by Engineering Research Center of Natural Enemies, Institute of Biological Control, Jilin Agricultural University, Changchun, China. In the laboratory, *T. dendrolimi* stock cultures were reared on eggs of *A. pernyi* as a factitious host. Rearing conditions were 27 °C ± 1, 75% ± 5 relative humidity (RH) and a 16:8 h (L:D) photoperiod.

2.2. Preparation of Artificial Medium and Insect Rearing

The artificial medium used in this study was the modified artificial medium developed by Lü et al. [9]. It comprised 3 mL of the pupal hemolymph of *A. pernyi*, 2.5 mL egg yolk, 1 mL 10% malted milk solution, 1 mL Neisenheimer's salt solution, 0.1 g trehalose (Sigma, St. Louis, MO, USA), and 1.5 mL sterile water. The preparation of artificial egg cards was done as described by Lü et al. [11].

Artificial egg cards were placed in a plastic tray (20 cm × 10 cm × 3 cm) for 24 h exposure to *T. dendrolimi* adults of the same batch. Parasitoids of both sexes were released in the trays using a 6:1 ratio of parasitoids to artificial eggs. Sex ratios were approximately 8:1 (female:male) in all three replicates (one tray corresponded to one replicate). Trays were placed in climatic incubators (Yamato, Tokyo, Japan) set at 27 °C ± 1, 75% ± 5 RH, and a 16:8 h (L:D) photoperiod. After 24 h of exposure, the wasps were removed, and the egg cards were transferred to the temperature treatments [9].

2.3. Experimental Set-Up, Sample Collection, and Biological Parameters Assessment

In a pre-experiment, in vitro reared *T. dendrolimi* were unable to complete the entire development (from egg to adult) at 10 °C, and stopped developing at prepupal or pupal stages. The optimum storage condition for these parasitoid wasps are prepupae that can be stored at 13 °C for up to 4 weeks without affecting reproductive quality [12]. The present experiment had two factors: temperature (13 °C ± 1 (optimum storage temperature), 16 °C ± 1 (above the optimum storage temperature) and 27 °C ± 1 (optimum development temperature)), and developmental stage (larva, prepupa, pupa, and adult). In a preliminary experiment, based on the transparency of the egg cards, they can be easily monitored daily for parasitoid development using a binocular microscope. *T. dendrolimi* developing to new larvae, prepupae, pupae, or adults were collected on ice and maintained at −80 °C for gene expression and biochemical assessments.

In the follow-up rearing experiment, to investigate the developmental quality of *T. dendrolimi* reared on artificial medium at different temperatures (27 °C ± 1, 23 °C ± 1, 20 °C ± 1, 16 °C ± 1, and 13 °C ± 1), the developmental durations (from oviposition to adult emergence) of the eggs, larvae, prepupae, and pupae, number of male adults and total adults observed per egg card were examined. Pupation rate, adult emergence rate (based on pupal numbers), numbers of normal adults produced (i.e., adults not having an enlarged abdomen and/or unexpanded wings) and the adult sex ratio (female proportion) was calculated as follows:

- Pupation rate (%) = (number of pupae/total number of larvae observed per egg card) × 100.
- Emergence rate (%) = (number of adults/(number of adults + dead pupae + dead larvae)) × 100.
- Number of normal adults = total number of normal adults observed to emerge from three replicates (egg cards)/3.
- Female proportion (%) = (total number of adults observed − number of male adults observed)/total number of adults observed × 100.

2.4. Total RNA Extraction and Cloning of the Full-Length cDNA

Total RNA was extracted from *T. dendrolimi* adults using a TransZol Up Plus RNA Kit (TransGen, Beijing, China). The RNA integrity and concentration were checked by agarose gel electrophoresis and spectrophotometry (NanoDrop2000, Wilmington, DE, USA), respectively. The fragments of *TdTPS* and *TdTRE* were obtained by transcriptome sequencing of *T. dendrolimi* (Hiseq 2000, Illumina, Beijing, China). Full-length sequences of TPS and TRE were obtained by 5′ and 3′ rapid amplification of cDNA ends (RACE) with the SMART™ RACE Kit (TaKaRa, Tokyo, Japan), according to manufacturer instructions. 5′ and 3′ RACE were performed by nested PCR including Universal Primer Mix (UPM) and Nested Universal Primer (NUP) along with gene-specific primers (GSP) (Table 1). The PCR conditions were as follows: initial at 94 °C for 5 min, 32 cycles of 30 s at 94 °C, 30 s at 60 °C, 2 min at 72 °C, and a final extension at 72 °C for 10 min. The products were examined by agarose gel electrophoresis, purified using a SanPrep Column DNA Gel Extraction Kit (Sangon, Shanghai, China), ligated into a Pucm-T vector (Sangon, Shanghai, China), and sequenced by Sanger's method.

Table 1. Primers sequences used for real-time PCR.

Gene	Primer Name	Sequence (5′-3′)	Usage
TdTPS	5Race-R	TTTTTTTTTTTTTTTTTTTTTTTTTTTVN	First full length cDNA synthesis for 5′cDNA amplification
	UPM-long	CTAATACGACTCACTATAGGGCAAGCAGTGGTATCAACGCAGAGT	
	UPM-short	CTAATACGACTCACTATAGGGC	
	BD SMART II™ A Oligonucleotide	AAGCAGTGGTATCAACGCAGAGTACGCGGG	
	3′NUP-R	AAGCAGTGGTATCAACGCAGAGT	
	3Race-R	AAGCAGTGGTATCAACGCAGAGTACT(30)VN	First full length cDNA synthesis for 3′cDNA or CDS amplification
	5r-TPS-1R	AGTGGTCCGCGATGAAGGTCGC	5′ cDNA amplification
	5r-TPS-2R	CCCAGATGCCATTTCCATTGATGAC	
	cTPS-F	TCGGGCAGYATGATYGTCGT	cds amplification
	cTPS-R	TGCCTCTCKACCCAYTTGAGCAT	
	cTPS-1R	TGTCTGTTACGTGGTCTGGGTGA	
	3r-TPS-R	CCACCACGACAACTCCTCGA	3′ cDNA amplification
	qTPS-F	AATGGAAATGGCATCTGGGTC	qRT-PCR
	qTPS-R	AGCAGCCGTTGTAGTACGAGTC	
TdTRE	5r-TRE-1R	GTACAGCTCGTCCTTGTCATCG	5′ cDNA amplification
	5r-TRE-2R	CGCTAATTGTATCGTCTTTAGTAGTTCG	
	cTRE-F	TTCCTGAAACAGTAGTMTTTAGTCG	cds amplification
	cTRE-R	CTCAAAAGTACGTTGTCCAAATAGAT	
	3r-TRE-R	GTTGATATCAAGAAACCAACGAACG	3′ cDNA amplification
	qTRE-F	AAGCGAAAGCCAAGCAAGGT	qRT-PCR
	qTRE-R	TGATACACGGGGTCACGAATAC	

UPM, Universal Primer Mix; NUP, Nested Universal Primer; F, Forward; R, Reverse.

2.5. Sequence Analysis

The amino acid sequences of *T. dendrolimi* in the fasta format was used to query the sequence database of the National Center for Biotechnology Information (NCBI) to identify proteins with primary sequence similarity to *TdTPS* and *TdTRE*. Multiple sequence alignment was constructed using MEGA 7 [48] with the CLUSTAL V method [49,50]. Phylogenetic trees were constructed using the neighbor-joining (NJ) method [51]. *D. melanogaster* was used as the out-group, and the stability of the tree was assessed via bootstrapping with >1000 replicates.

2.6. Expression of TdTPS and TdTRE

Total RNA was extracted from 0.1 g of *T. dendrolimi* at larval, prepupal, pupal, and adult stages using a TransZol Up Plus RNA Kit (TransGen, Beijing, China). First-strand cDNA was synthesized from 1 μg total RNA using a FastQuant RT Kit With gDNase (Tiangen, Beijing, China).

Steps to construct linearized plasmid standards were as described previously with some modifications [52]. First, products *TdTPS* and *TdTRE* were extracted and purified with an agarose gel Extraction Kit (Sangon, Shanghai, China). Second, each gene was cloned separately using the Pucm-T Cloning Vector Kit (Sangon), according to the manufacturer instructions. Third, positive clones screened by PCR were processed for plasmid isolation using a Plasmid Extraction & Purification Kit (Sangon) and confirmed by Sanger sequencing (Sangon). Fourth, plasmids were completely linearized by *EcoR* I digestion for 4.5 h at 37 °C and confirmed by checking band patterns in the agarose gel. Fifth, linearized plasmids were quantified using a NanoDrop2000 spectrophotometer (NanoDrop), and copy numbers were calculated for all standards by the following formula [53]:

$$\text{Copies}/\mu\text{L} = \frac{\left(6.02 \times 10^{23}\ \text{copies}\right) \times \left(\text{plasmid concentration g}/\mu\text{L}\right)}{\left(\text{number of bases}\right) \times \left(660\ \text{daltons}/\text{base}\right)}$$

Finally, the standard DNA (template) was prepared in a dilution series from 10^{-3} to 10^{-10} (copies/5 μL) for qPCR. qPCR was performed using a SYBR Green Mix Kit (Tiangen, Beijing, China) to measure the cycle number (Ct) of each dilution in duplicate. Each PCR reaction was mixed with 10 μL SYBR Green Mix, 6 μL dd H_2O, 2 μL cDNA, and 1 μL of each primer (10 μM). Cycling conditions for all standards were described as above followed with dissociation curve analysis. Standard curves were generated as linear regression between Ct and log10 starting copy number of standard DNA. The Ct values were reported by the MX3000P MXPro program (Agilent Technologies, Palo Alto, USA). Amplification efficiency, slopes, and correlation coefficient (R^2) were automatically calculated by the program.

To study the gene expression profiles during the four life stages, absolute quantitative PCR (AQ-PCR) was conducted to estimate their starting copy numbers. Gene specific primers, qTdTPSF/R, and qTdTREF/R, were designed according to the full-length cDNAs and these are listed in Table 1. RT-PCR was performed to obtain gene targets in the following cycling condition: initial denaturation 3 min at 95 °C followed by 40 cycles including 5 s at 95 °C, 10 s at 55 °C, 15 s at 72 °C. In the case of the *T. dendrolimi* samples, qPCR Ct values of *TdTPS* and *TdTRE* expression profiles were used to estimate starting copy numbers based on their standard curves.

2.7. Enzyme Activity Measurements

To obtain crude extracts for enzyme activity study, each sample (0.0100 ± 0.0002 g of larvae, pupae, prepupae, and adults) was homogenized at 0 °C (TGrinder OSE-Y20 Homogenizer, Tiangen, Beijing, China) after adding 2000 μL of 20 mM phosphate buffered saline (PBS, pH 5.8). The homogenates were centrifuged at 10,000× *g* at 4 °C for 20 min (CP100MX, Hitachi, Tokyo, Japan), and cuticle debris was removed. Supernatants in PBS were maintained at −80 °C to analyze TPS and TRE activity.

To determine TPS activity, the method of Dual et al. [54] was used. The qualitative analysis of trehalose was performed by thin layer chromatography (TLC); the quantitative analysis of trehalose synthase activity was investigated by examining the difference in glucose that was formed from the hydrolysis of maltose by TPS in the presence and absence of α-glucosidase and measured by DNS. The reaction mixture, containing 150 μL of crude extract and 100 μL substrate (10% maltose) was inactivated in a water bath at 60 °C for 1 h, and then in boiling water bath for 10 min. After adding 85% phosphoric acid to adjust pH to 4.2, 1 mL of diluted maltase (alpha-glucosidase) was added to the mixtures. The mixtures were inactivated in a water bath at 60 °C for 1 h and then in a boiling water bath for 10 min. Diluting and volumetrizing the reaction solution, adding 1:2 (V reaction solution: V DNS method), boiling water bath for 10 min, cooled in an ice bath, then the absorbance was measured at 550 nm.

TRE activity was measured as described previously by the 3,5-dinitrosalicylic acid colorimetric method (DNS method) with absorbance measured at 540 nm [55,56]. One unit (U) of enzyme activity was defined as the amount of enzyme capable of releasing 1 mg of reducing sugar per minute. The reaction mixture consisted of 500 μL crude extract and 500 μL DNS solution. The reaction was stopped by heating in boiling water for 5 min, then 4 mL of a pH 5.8 KH2PO4-NAOH buffer solution was added.

2.8. Statistical Analyses

Each treatment was performed using three biological and three technical replicates. Multifactor analysis (PROC GLM) of variance was conducted to evaluate temperatures and effect of developmental stage on the gene expression and enzyme activities of parasitoids using Tukey's test. The qualities of biological parameters were compared using one-way analysis of variance (ANOVA) and multiple comparisons of means were conducted using Tukey's test. Before analysis, percentage data were arcsine square root–transformed, and the data on the number of adults produced were $\log_{10}$-transformed to fit a normal distribution. For absolute quantification analysis, the number of molecules was expressed as the mean of the $\log_{10}$-converted value $\pm$ standard error. In all experiments, differences among means were considered significant at $p < 0.05$. Statistical analyses were conducted using SPSS 22.0 software (SPSS Inc., Chicago, IL, USA).

3. Results

3.1. Cloning and Characterization of Full-Length TdTPS and TdTRE cDNAs

To identify the *TdTPS* and *TdTRE*, cDNA fragments involved in trehalose metabolism were identified through *T. dendrolimi* transcriptome data (Hiseq 2000, Illumina, Santiago, USA). Based on the cDNA fragments, specific primers were performed and full-length cDNA of *TdTPS* and *TdTRE* was obtained by RACE-PCR. The full length *TdTPS* gene has 3189 bp, and the cDNA has a 2358 bp open reading frame (ORF), encoding a polypeptide of 785 amino acids with an estimated molecular weight of approximately 88.60 kDa and a pI of 6.56. The ORF was identical to the homolog from *Trichogramma pretiosum* (100%) (Table 2). Sequence analysis showed that the deduced amino acid sequence includes a conserved TPS domain (aa 1–478, E = 1×10^{-133}) and a TPP domain (aa 515–739, E = 7×10^{-34}). Multiple protein alignment indicated that the *TdTPS* protein contains six conserved motifs (Figure 1). In addition to these specific motifs, *TdTPS* proteins contain the highly conserved domains, HDYHLML and DGMNLV, the same as TPS genes previously reported, such as in *D. melanogaster*, *Catantops pinguis*, and *Delia antiqua* [20,34,41]. Phylogenetic analysis showed that the *TdTPS* was more closely related to those of other Hymenoptera species (*T. pretiosum*, *Nasonia vitripennis*, and *Copidosoma floridanum*), and could be assigned to the same subgroup (Figure 2). The cloned TPS gene was designated as *TdTPS* and deposited into GenBank (MT108781).

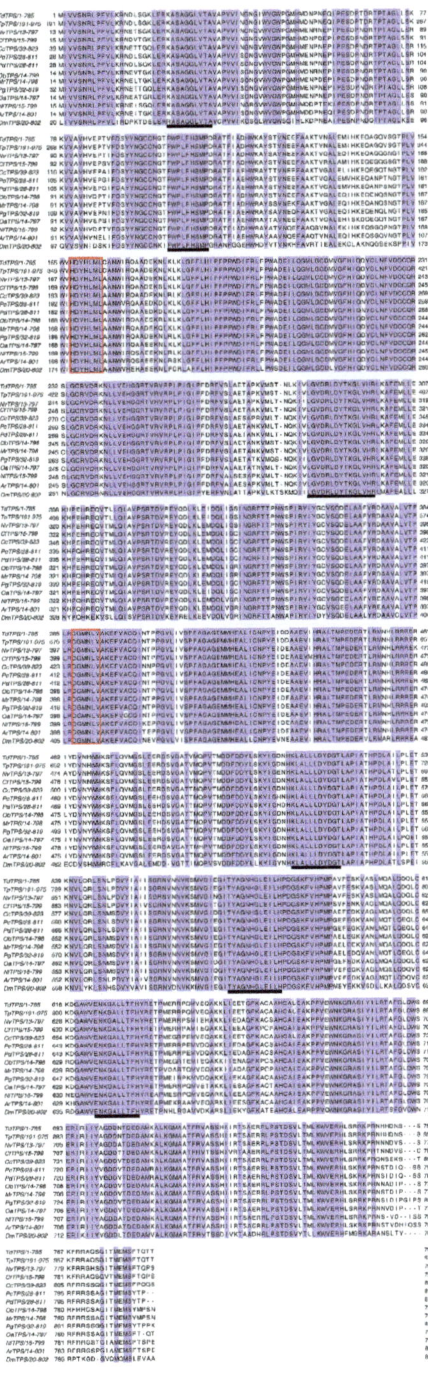

Figure 1. Alignment of *TdTPS*. Multiple alignment of trehalose-6-phosphate synthase (TPS) protein sequences from different insect species. Identical amino acid residues are shown in purple. The conserved motifs and the signatures are underlined and boxed, respectively.

Table 2. The identities of *TdTPS* and *TdTRE* genes from different insects with *T. dendrolimi*.

Genes	Insects	GenBank Number	Identity
TdTPS	*Trichogramma pretiosum*	XP_014221069	100%
	Nasonia vitripennis	XP_016837588	94.78%
	Copidosoma floridanum	XP_014213166	93.89%
	Cephus cinctus	XP_015588847	92.87%
	Osmia bicornis	XP_029055554	92.45%,
	Megachile rotundata	XP_003702415	92.32%
	Polistes canadensis	XP_014609582	92.21%
	Polistes dominula	XP_015172546	92.08%
	Orussus abietinus	XP_012281922	91.96%
	Pseudomyrmex gracilis	XP_020289281	91.85%
	Neodiprion lecontei	XP_015522281	91.30%
	Athalia rosae	XP_012252443	90.83%
TdTRE	*Trichogramma pretiosum*	XP_014236786	97.61%
	Trichomalopsis sarcophagae	OXU30694	67.52%
	Ceratosolen solmsi marchali	XP_011497766	67.18%
	Nasonia vitripennis	XP_008215783	65.94%
	Copidosoma floridanum	XP_014216724	60.41%
	Harpegnathos saltator	XP_011144292	57.38%
	Solenopsis invicta	XP_011170317	56.69%
	Monomorium pharaonic	XP_028048276	56.42%
	Pseudomyrmex gracilis	XP_020280302	56.31%
	Orussus abietinus	XP_012271873	55.95%

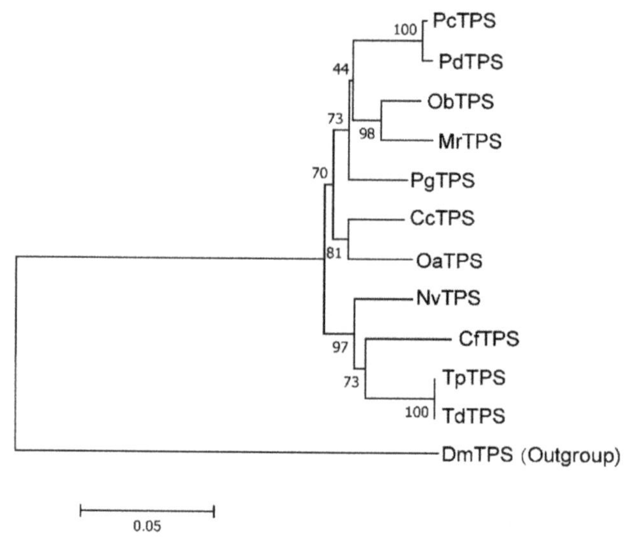

Figure 2. Phylogenetics of *TdTPS*. Phylogenetic tree constructed using the neighbor-joining (NJ) method. Percentage bootstrap values larger than 40 are shown on each branch. *PcTPS: Polistes canadensis*, XP_014609582; *PdTPS: Polistes dominula*, XP_015172546; *ObTPS: Osmia bicornis*, XP_029055554; *MrTPS: Megachile rotundata*, XP_003702415; *PgTPS: Pseudomyrmex gracilis*, XP_020289281; *CcTPS: Cephus cinctus*, XP_015588847; *OaTPS: Orussus abietinus*, XP_012281922; *NvTPS: Nasonia vitripennis*, XP_016837588; *CfTPS: Copidosoma floridanum*, XP_014213166; *TpTPS: Trichogramma pretiosum*, XP_014221069; *DmTPS: Drosophila melanogaster*, ABH06641.1.

The full length *TdTRE* cDNA consisted of 2228 bp, including an 1878 bp open reading frame, encoding 625 amino acids with a predicted molecular weight of 73.2 kDa and a pI of 6.40. Basic

Local Alignment Search Tool (BLAST) analysis revealed that *TdTRE* is 55.95–97.61% identical in structure to other known insect TRE forms. *TdTRE* is also most similar to the TRE from *T. pretiosum* (97.61%) (Table 2). The deduced amino acid sequence of *TdTRE* contains one conserved motif (YYLMRSQPPLLIPM) and a signal peptide sequence (Figure 3). Moreover, *TdTRE* had the same signature motifs, PGGRFREFYYWDSY and QWDYPNAWPP. For phylogenetic analysis, *TdTRE* was clustered with *T. pretiosum* TRE (Figure 4). Based on the sequence identity with known soluble form trehalase genes, *TdTRE* was identified as a soluble trehalase and deposited into GenBank (MT108782).

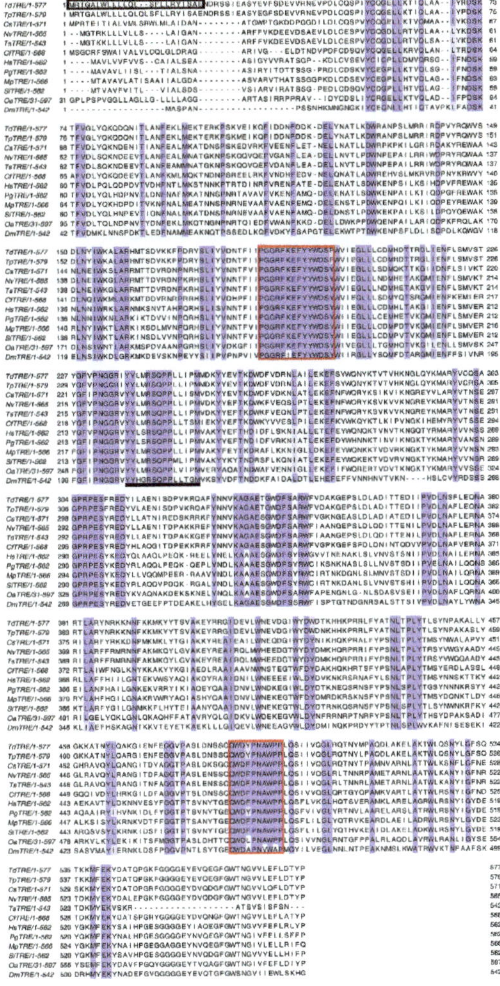

Figure 3. Alignment of *TdTRE*. Multiple alignment of soluble trehalase (TRE) protein sequences from different insect species. Identical amino acid residues are shown in purple. The conserved motifs, the signatures, and signal peptide sequence are underlined, red boxed, and black boxed, respectively.

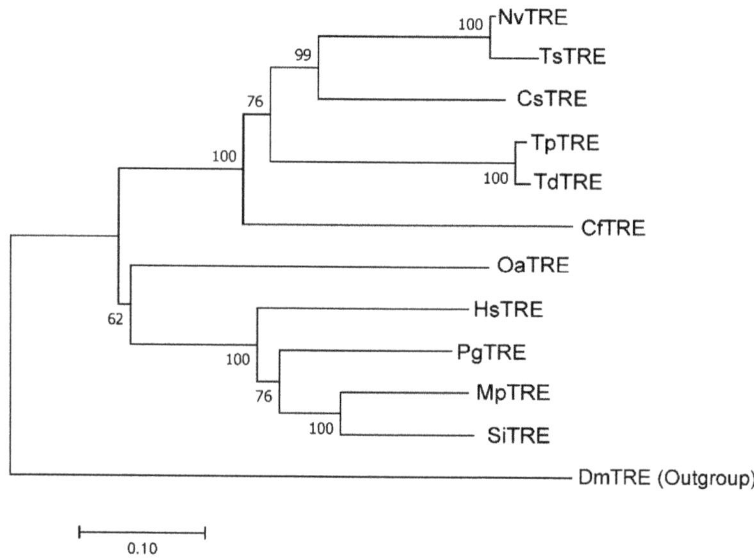

Figure 4. Phylogenetics of *TdTRE*. Phylogenetic tree constructed using the neighbor-joining (NJ) method. Percentage bootstrap values larger than 40 are shown on each branch. *NvTRE: Nasonia vitripennis*, XP_008215783; *TsTRE: Trichomalopsis sarcophagae*, OXU30694; *CsTRE: Ceratosolen solmsi marchali*, XP_011497766; *TpTRE: Trichogramma pretiosum*, XP_014236786; *CfTRE: Copidosoma floridanum*, XP_014216724; *OaTRE: Orussus abietinus*, XP_012271873; *HsTRE: Harpegnathos saltator*, XP_011144292; *PgTRE: Pseudomyrmex gracilis*, XP_020280302; *MpTRE: Monomorium pharaonis*, XP_028048276; *SiTRE: Solenopsis invicta*, XP_011170317; *DmTRE: Drosophila melanogaster*, NP_726025.1.

3.2. Standard Curve

Absolute quantification determines the actual copy numbers of target genes (*TdTPS* and *TdTRE*) by relating the Ct value to a standard curve and amplifying serial dilutions of plasmid standards by qPCR. The Ct values were measured and plotted against known copy numbers of the standard sample. The reaction efficiency and linearity for the serially diluted standards were of good quality for both genes (Figure 5). The standard curve covered a linear range of seven orders of magnitude. The slope (-3.3550 and -3.3396) and the correlation coefficient ($R^2 = 0.9992$ and 0.9990) of the standard curve indicated that this assay could be used to quantify target RNA in *T. dendrolimi*.

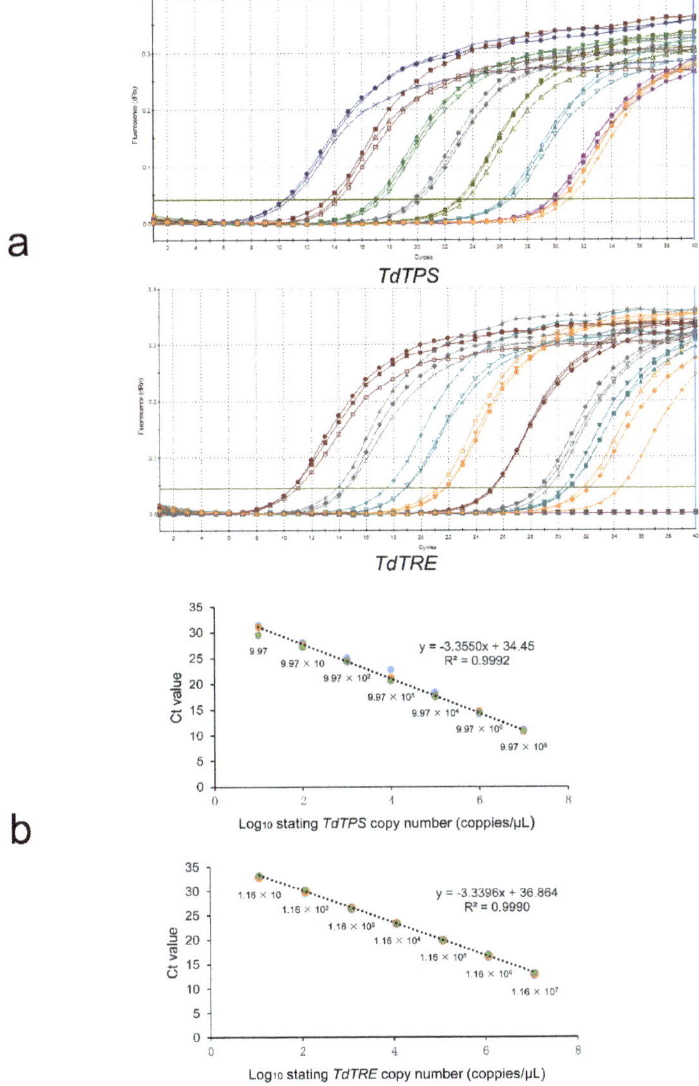

Figure 5. Standard curve for RT-qPCR amplification of standard sample. (**a**) Amplification plots for *TdTPS* and *TdTRE*; (**b**) standard curves of real-time PCR of *TdTPS* and *TdTRE*, using the method of absolute quantitative analysis, showing the testing in triplicate of a 10-fold dilution series containing a standard sample ranging from 9.97×10^6 to 9.97 and 1.16×10^7 to 1.16×10^1 copies per.

3.3. Effect of Temperature on the Expression of TdTPS and TdTRE during Development

Absolute quantification PCR (AQ-PCR) experiments to measure *TdTPS* and *TdTRE* absolute expression in four developmental stages at different temperatures revealed that there was interaction between two factors on the mRNA levels of the genes (Table 3). Table 4 shows that the levels of the two genes were more highly expressed at the optimum storage temperature (13 °C) in all developmental stages compared to the expression at the other treatment temperatures. The *TdTRE* transcripts were highly expressed at 16 °C. In contrast, at the optimum development temperature (27 °C), the expression levels of *TdTPS* and *TdTRE* were low in all developmental stages. *TdTPS* was highly expressed in prepupae when *T. dendrolimi* developed at the three treatment temperatures, and *TdTPS* expression level was also higher in adults when they developed at 16 °C. For trehalase, the absolute expression levels of *TdTRE* were very low in the larval stage at all temperatures, and much higher in the prepupal stage at 27 °C and 16 °C.

Table 3. Multifactor variance analysis of effects of two factors on gene absolute expression and enzymes activity of *Trichogramma dendrolimi* reared on artificial medium.

Parameters	Factors	df	F	p
Absolute	A	2	2348.95	<0.001
expression of	B	3	107.13	<0.001
TdTPS	A × B	6	29.89	<0.001
Absolute	A	2	462.30	<0.001
expression of	B	3	337.41	<0.001
TdTRE	A × B	6	8.25	<0.001
	A	2	25.24	<0.001
Activity of *TdTPS*	B	3	18.12	<0.001
	A × B	6	1.41	0.252
	A	2	7.10	<0.001
Activity of *TdTRE*	B	3	39.86	<0.001
	A × B	6	46.64	<0.001

A, temperature; B, developmental stage.

Table 4. The gene expression of *TdTPS* and *TdTRE* of in vitro-reared *Trichogramma dendrolimi* reared at different temperatures and developmental stages.

Gene	Developmental Stage	Gene Absolute Expression (Copy Number (Coppies/μL))		
		13 °C	16 °C	27 °C
TdTPS	Larva	46175.06 ± 6494.615 A c	1419.43 ± 109.256 B b	1082.30 ± 53.480 B b
	Prepupa	221227.25 ± 8108.441 A a	9516.15 ± 473.110 B a	1809.35 ± 72.933 B a
	Pupa	84851.80 ± 5186.980 A b	4581.10 ± 1153.243 B b	859.34 ± 20.148 B bc
	Adult	66806.48 ± 1349.493 A bc	9716.46 ± 799.299 B a	774.88 ± 37.565 C c
TdTRE	Larva	8.36 ± 0.933 A c	9.72 ± 0.758 A b	2.41 ± 0.566 B c
	Prepupa	126.08 ± 7.207 A b	158.51 ± 10.949 A a	17.78 ± 1.506 B a
	Pupa	122.12 ± 4.192 A b	136.25 ± 7.018 A a	11.48 ± 2.241 B ab
	Adult	162.04 ± 12.807 A a	129.57 ± 3.250 A a	11.19 ± 0.886 B b

Mean ± SE were calculated from three replicates. Mean ± SE followed by the same capital letter within a row were not significantly different (Tukey's test: $p > 0.05$); Mean ± SE followed by the same lowercase letter within a column were not significantly different (Tukey's test: $p > 0.05$).

3.4. Changes in Enzyme Activities

TdTPS and *TdTRE* enzyme activities were compared at different temperatures and different developmental stages, respectively. Comparison among temperatures indicated that *TdTPS* activities were similar at normal and cold temperatures. In larval and adult stages, *TdTRE* showed much higher activities at 16 °C and 13 °C. However, the *TdTRE* activity in prepupae showed the opposite result. In the pupal stage, *TdTRE* activities were similar at the three temperatures (Tables 3 and 5).

At 27 °C, *TdTPS* activity in the developmental stages was similar. *TdTRE* had significantly higher activity in the prepupal and pupal stages and then declined in the adult stage. The enzyme activity of *TdTPS* and *TdTRE* had a similar trend at 16 °C and 13 °C in all *T. dendrolimi* developmental stages. Enzyme activities of *TdTRE* in prepupae declined sharply when they developed at 13 °C and 16° C.

Table 5. Activity of the enzymes involved in trehalose metabolism of in vitro-reared *Trichogramma dendrolimi* reared at different temperatures and developmental stages.

Enzymes	Developmental Stage	Activity [µg/mL (Extract/min)]		
		13 °C	**16 °C**	**27 °C**
TdTPS	Larva	0.15 ± 0.009 A ab	0.14 ± 0.012 A b	0.09 ± 0.014 A a
	Prepupa	0.12 ± 0.005 A b	0.11 ± 0.009 A b	0.08 ± 0.021 A a
	Pupa	0.12 ± 0.015 A b	0.11 ± 0.006 A b	0.09 ± 0.006 A a
	Adult	0.20 ± 0.014 A a	0.19 ± 0.008 A a	0.11 ± 0.010 B a
TdTRE	Larva	0.52 ± 0.032 A a	0.44 ± 0.028 A a	0.28 ± 0.022 B b
	Prepupa	0.22 ± 0.032 B b	0.18 ± 0.016 B b	0.49 ± 0.021 A a
	Pupa	0.47 ± 0.019 A a	0.50 ± 0.031 A a	0.50 ± 0.012 A a
	Adult	0.50 ± 0.013 A a	0.50 ± 0.002 A a	0.22 ± 0.012 B b

Mean ± SE were calculated from three replicates. Mean ± SE followed by the same capital letter within a row were not significantly different (Tukey's test: $p > 0.05$); Mean ± SE followed by the same lowercase letter within a column were not significantly different (Tukey's test: $p > 0.05$).

3.5. In Vitro Rearing at Different Temperatures

The developmental durations and biological parameters at the five temperatures are shown in Table 6. The development of *T. dendrolimi* at 13 °C, 16 °C, 20 °C, and 23 °C prolonged for 21 days, 17 days, 8 days, and 5 days, respectively compared with those reared at the optimum temperature (27 °C). There were no significant differences in pupation rate, emergence rate, female proportion and number of normal adults among 16 °C, 20 °C, 23 °C, and 27 °C ($F_{4,14}$ = 4.070, 135.396, 2.430, and 64.434, p = 0.033, 0.000, 0.116 and 0.000, respectively). The biological parameters (except female proportion) of *T. dendrolimi* reared on artificial medium were significantly affected by temperature. These parameters at 13 °C was lowest compared to that of other test temperatures.

Table 6. Developmental quality of in vitro reared *T. dendrolimi* at different temperatures.

Temperature	Developmental Duration (D)					Biological Parameters			
	Egg	Larva	Prepupa	Pupa	Total Duration	Pupation Rate (%)	Emergence Rate (%)	Female Proportion (%)	Number of Normal Adults
27 °C	2	2	2	4	10	96.77 ± 0.391 a	77.44 ± 1.451 a	88.16 ± 0.555 a	876.00 ± 48.03 a
23 °C	3	3	2	5	13	96.44 ± 0.441 ab	75.16 ± 1.326 a	87.50 ± 0.620 a	805.33 ± 35.044 a
20 °C	4	4	3	7	18	96.48 ± 0.395 ab	76.59 ± 0.728 a	87.62 ± 0.949 a	814.67 ± 13.119 a
16 °C	5	9	4	9	27	94.88 ± 1.35 ab	75.86 ± 2.499 a	88.00 ± 0.392 a	904.00 ±72.746 a
13 °C	6	9	7	9	31	92.81 ± 1.041 b	36.03 ± 1.159 b	85.55 ± 0.724 a	92.67 ± 5.897 b

Mean ± SE values were calculated from three replicates. Mean ± SE followed by the same lower case letter within a column are not significantly different (Tukey's test: $p < 0.05$).

4. Discussion

In this study, only the TPS gene of *T. dendrolimi* was obtained. The deduced amino acid sequence reveals that *TdTPS*, similar to the *DaTPS* gene in *Delia antiqua* [20], has two conserved functional domains that include an N-terminal TPS domain and a C-terminal TPP domain [43]. This result supports the conclusion that the insect TPS is a fused gene [57]. There are two signature motifs (HDYHL and DGMNLV) of TPS protein sequences for insects, plants, bacteria, fungi, and nematodes. Multiple protein alignment results show that, besides the signature motifs, *TdTPS* has other conserved motifs. Two types of trehalase exist in insects, a soluble trehalase and a membrane-bound trehalase with a transmembrane domain near the C-terminus [2,29,44,58,59]. Based on the conserved motif and specific signatures in deduced amino acid sequences, we identified the existence of the soluble form of trehalase (*TdTRE*) in in vitro reared *T. dendrolimi*.

Trehalose occurs in all insect species and is the most characteristic sugar in the insect hemolymph [57]. However, it has not been detected in the developmental stages of some species [2,23]. In the present study, *TdTPS* and *TdTRE* were expressed in larval, prepupal, pupal, and adult stages, whether they developed at the optimum development temperature or at sustained low temperatures.

The results suggest that sustained low temperature has a strong effect on the expression level of trehalase genes in *T. dendrolimi*, which may facilitate the utilization of trehalose. A comparison of the expression levels of *TdTPS* and *TdTRE* at the tested temperature showed that sustained low temperature can upregulate their expression, but the expression of *TdTRE* was much lower than *TdTPS*. These indicated that the anabolism of trehalose was greater than its catabolism during low temperature development. Meanwhile, the reduction of trehalase activity inhibits the trehalose catabolism, which indirectly helps the trehalose accumulation. *TdTPS* might play a more important role than *TdTRE* in cold induction.

Prepupal stage is a critical developmental period for *Trichogramma*. The nutrient (host egg liquid) has been consumed by the end of larval stage and digestion of this food allows for the accumulation of energy used by the following pupal and adult stages. There were studies that have reported that the prepupa is the best stage for short-term storage of *Trichogramma* spp. reared in vivo [12]. In a previous study, sustained temperatures at 13 °C for 4 weeks was the optimum short-term storage condition for prepupae of in vitro reared *T. dendrolimi* [12]. Here, we explain the results of the previous study in terms of molecular physiology. *TdTPS* had a higher expression level in prepupae at the three test temperatures. Meanwhile, *TdTPS* had a high expression level in all stages when *T. dendrolimi* developed at low temperatures, especially at 13 °C. The enzymatic activity of *TdTPS* showed a similar change during the entire development. With a decrease of temperature, the activity increased. *TdTRE* showed a high expression level, but the enzymatic activity decreased at low temperatures in the prepupal stage. This trend was opposite to that observed at 27 °C. These findings indicate that the gene regulation for *TdTRE*, the soluble trehalase, might not determine the enzyme activity directly. However, there is no evidence to prove whether other soluble trehalase or membrane-bound trehalase genes exist in *T. dendrolimi* yet. Large-scale expression of TPS by insects before pupation promotes the synthesis of trehalose, leading to a high level of trehalose in the pupa stage. Under sustained low temperature, trehalase activity was inhibited in the prepupae since trehalose accumulation was probably required during this period to meet the energy required for chitin synthesis in the pupa and for adult emergence, suggesting its potential role in molting from pre-pupae to pupae. However, *TdTRE* is more active than *TdTPS* during cold stress. In vitro reared *T. dendrolimi* does not diapause under sustained low temperature stress, but cold only prolongs the development period after which development proceeds normally. In addition, the trehalose supplemented in artificial medium [9] needs to be consumed. On the other hand, in vitro reared *T. dendrolimi* could complete development above 13 °C because the trehalose added to the artificial medium may improve its cold tolerance. This protected the parasitoids, especially the young larvae, from low temperature injury. Trehalose was synthesized and accumulated at the same time for metabolism and utilization during the period. This indicates that trehalose might regulate growth of in vitro reared *T. dendrolimi* and the metabolic process of cold tolerance.

Trehalose concentration in insect blood hemolymph is not under homeostatic regulation. It is based on environmental conditions, physiological state, and nutrition. *Trichogramma* complete their development in the host egg before adult emergence. During development, the accumulation of trehalose helps the insects to resist environmental temperature stress and the stress of host malnutrition. A comparison of the trehalose contents and trehalase activity of *T. dendrolimi* produced in vitro and in vivo showed that the adults produced in vitro had higher trehalose content and trehalase activity over 10 generations [11]. Therefore, the responses and resistance of egg parasitoids to environmental stress may be different from those of other insects.

Future work will need to focus on the effects of different hosts (nutrition) on trehalose metabolism in *T. dendrolimi*, and determine how the trehalase genes are regulated under stress conditions.

5. Conclusions

Trehalose synthetase and soluble trehalase genes were identified from *T. dendrolimi* reared on an artificial medium. Sustained low temperature stress had different effects on trehalose metabolism related enzyme genes and enzyme activities of *Trichogramma*. The anabolism and catabolism of trehalose maintained a dynamic balance in the process of metabolism. Trehalose indeed accumulated as an energy source to be used in adverse conditions. *TdTPS* and *TdTRE* may be considered as an energy source and responsive enzymes for cold resistance in *Trichogramma*. The prepupa stage is a key period of *Trichogramma* development, in which the expression of genes involved in trehalose metabolism and corresponding enzyme activities undergo substantial changes. It' is found that 13 °C appears to be the cold tolerance threshold temperature for in vitro reared *T. dendrolimi*. Since it remains unknown whether or not the in vitro reared *T. dendrolimi* could diapause without negatively affecting their reproductive parameters, we suggest that in vitro reared *T. dendrolimi* could be reared at temperatures of 16 °C, 20 °C, and 23 °C to reduce rearing costs.

Author Contributions: Conceptualization, X.L., J.L., and L.-y.L.; methodology, X.L.; formal analysis, X.L.; resources, S.-c.H. and Z.-g.L.; data curation, X.L.; writing—original draft preparation, X.L.; writing—review and editing, J.L., S.-c.H., and L.-y.L.; visualization, X.L. and Z.-g.L.; supervision, L.-y.L. and J.L.; project administration, X.L. and J.L.; funding acquisition, X.L., J.L., S.-c.H., and Z.-g.L. All authors have read and agreed to the published version of the manuscript.

Funding: This research was funded by the Key-Area Research and Development Program of Guangdong Province (grant no. 2020B020223004), GDAS Special Project of Science and Technology Development (grant nos. 2020GDASYL-20200301003, 2020GDASYL-20200104025, and 2018GDASCX−0107), National Natural Science Foundation of China (grant no. 31501702), Science & Technology Planning Project of Guangdong (grant no. 2019B030316018).

Acknowledgments: We would like to thank Patrick De Clercq provided guidance; Chuan-bu Gao prepared the figures; Jun-gang Qian assisted with the experiment.

Conflicts of Interest: The authors declare no conflict of interest.

References

1. Rivero, A.; Casas, J. Incorporating physiology into parasitoid behavioral ecology: The allocation of nutritional resources. *Res. Popul. Ecol.* **1999**, *41*, 39–45. [CrossRef]
2. Thompson, S.N. Trehalose—The insect 'blood' sugar. *Adv. Insect Physiol.* **2003**, *31*, 203–285. [CrossRef]
3. Phillips, C.B.; Hiszczynska-Sawicka, E.; Iline, I.I.; Novoselov, M.; Jiao, J.; Richards, N.K.; Hardwick, S. A modified enzymatic method for measuring insect sugars and the effect of storing samples in ethanol on subsequent trehalose measurements. *Biol. Control* **2008**, *126*, 127–135. [CrossRef]
4. Zhang, J.J.; Ruan, C.C.; Zang, L.S.; Shao, X.W.; Shi, S.S. Technological improvements for mass production of *Trichogramma* and current status of their applications for biological control on agricultural pests in China. *Chin. J. Biol. Control* **2015**, *31*, 638–646. [CrossRef]
5. Zhang, J.J.; Zhang, X.; Zang, L.S.; Du, W.M.; Hou, Y.Y.; Ruan, C.C.; Desneux, N. Advantages of diapause in *Trichogramma dendrolimi* mass production via eggs of the Chinese silkworm, *Antheraea Pernyi*. *Pest Manag. Sci.* **2018**, *74*, 959–965. [CrossRef] [PubMed]
6. Cônsoli, F.L.; Grenier, S. In vitro Rearing of Egg Parasitoids. In *Egg Parasitoids in Agroecosystems with Emphasis on Trichogramma*; Cônsoli, F.L., Parra, J.R.P., Zucchi, R.A., Eds.; Springer: Berlin/Heidelberg, Germany, 2010; pp. 293–313.
7. Li, L.Y.; Han, S.C.; Lü, X. Mass rearing *Trichogramma* on Artificial Media in China. In *Biological Control of Pests Using Trichogramma: Current Status and Perspectives*; Vinson, S.B., Greenberg, S.M., Liu, T.X., Rao, A., Volosciuc, L.F., Eds.; Northwest A & F University Press: Yangling, China, 2015; pp. 32–57.
8. Lü, X.; Han, S.C.; Li, L.Y.; Grenier, S.; De Clercq, P. The potential of trehalose to replace insect hemolymph in artificial media for *Trichogramma dendrolimi* Matsumura (Hymenoptera: Trichogrammatidae). *Insect Sci.* **2013**, *20*, 629–636. [CrossRef] [PubMed]
9. Lü, X.; Han, S.C.; De Clercq, P.; Dai, J.Q.; Li, L.Y. Orthogonal array design for optimization of an artificial medium for In vitro rearing of *Trichogramma dendrolimi*. *Entomol. Exp. Appl.* **2014**, *152*, 52–60. [CrossRef]
10. Lü, X.; Han, S.C.; Li, Z.G.; Li, L.Y. Biological characters of *Trichogramma dendrolimi* (Hymenoptera: Trichogrammatidae) reared in vitro versus in vivo for thirty generations. *Sci. Rep.* **2017**, *7*, 17928. [CrossRef]
11. Lü, X.; Han, S.-C.; Li, L.-Y. Biochemical analyses of *Trichogramma dendrolimi* (Hymenoptera: Trichogrammatidae) In vitro and in vivo rearing for 10 generations. *Fla. Entomol.* **2015**, *98*, 911–915. [CrossRef]

12. Lü, X.; Han, S.C.; Li, J.; Liu, J.S.; Li, Z.G. Effects of cold storage on the quality of *Trichogramma dendrolimi* Matsumura (Hymenoptera: Trichogrammatidae) reared on artificial medium. *Pest Manag. Sci.* **2019**, *75*, 1328–1338. [CrossRef] [PubMed]

13. Li, L.Y.; Zhang, Y.H.; Zhang, R.H. Influence of temperature on the growth and development of *Trichogramma* spp. on interspecific and intraspecific levels. *Nat. Enemies Insects* **1983**, *5*, 1–5.

14. Shi, Z.H.; Liu, S.S. The influence of temperature and humidity on the population increase of *Trichogramma dendrolimi*. *Acta Ecol. Sin.* **1993**, *13*, 328–333.

15. Wang, C.L.; Wang, H.X.; Wang, Y.A.; Lu, H. Studies on the relationship of temperature and development of *Trichogramma dendrolimi* Matsumura. *Zool. Res.* **1981**, *2*, 317–326.

16. Li, L.J.; Lu, X.; Zhang, G.H.; Zhou, S.X.; Chang, X.; Ding, Y. Difference of development characters of two *Trichogramma* Species at various temperatures. *J. Northeast Agr. Sci.* **2017**, *42*, 23–26. [CrossRef]

17. Hayakawa, Y.; Chino, H. Temperature-dependent activation or inactivation of glycogen phosphorylase and synthase of fat body of the silkworm, *Philosamia cynthia*: The possible mechanism of the temperature-dependent interconversion between glycogen and trehalose. *Insect Biochem.* **1982**, *12*, 361–366. [CrossRef]

18. Khani, A.; Moharramipour, S.; Barzegar, M. Cold tolerance and trehalose accumulation in overwintering larvae of the codling moth, *Cydia pomonella* (Lepidoptera: Tortricidae). *Eur. J. Entomol.* **2007**, *104*, 385. [CrossRef]

19. Tang, B.; Qin, Z.; Shi, Z.K.; Wang, S.; Guo, X.J.; Wang, S.G.; Zhang, F. Trehalase in *Harmonia axyridis* (Coleoptera: Coccinellidae): Effects on beetle locomotory activity and the correlation with trehalose metabolism under starvation conditions. *Appl. Entomol. Zool.* **2014**, *49*, 255–264. [CrossRef]

20. Guo, Q.; Hao, Y.J.; Li, Y.; Zhang, Y.J.; Ren, S.; Si, F.L.; Chen, B. Gene cloning, characterization and expression and enzymatic activities related to trehalose metabolism during diapause of the onion maggot *Delia antiqua* (Diptera: Anthomyiidae). *Gene* **2015**, *565*, 106–115. [CrossRef]

21. Tang, B.; Wei, P.; Zhao, L.N.; Shi, Z.K.; Shen, Q.D.; Yang, M.M.; Xie, G.Q.; Wang, S.G. Knockdown of five trehalase genes using RNA interference regulates the gene expression of the chitin biosynthesis pathways in *Tribolium castaneum*. *BMC Biotechnol.* **2016**, *16*, 67. [CrossRef]

22. Wingler, A. The function of trehalose biosynthesis in plants. *Phytochemistry* **2002**, *60*, 437–440. [CrossRef]

23. Elbein, A.D.; Pan, Y.T.; Pastuszak, I.; Carroll, D. New insights on trehalose: A multifunctional molecule. *Glycobiology* **2003**, *13*, 17R–27R. [CrossRef]

24. Heimpel, G.E.; Rosenheim, J.A.; Kattari, D. Adult feeding and lifetime reproductive success in the parasitoid *Aphytis melinus*. *Entomol. Exp. Appl.* **1997**, *83*, 305–315. [CrossRef]

25. Olson, D.; Andow, D. Larval crowding and adult nutrition effects on longevity and fecundity of female *Trichogramma nubilale* Ertle & Davis (Hymenoptera: Trichogrammatidae). *Environ. Entomol.* **1998**, *27*, 508–514. [CrossRef]

26. Wyatt, G.R. The biochemistry of sugars and polysaccharides in insects. *Adv. Insect Physiol.* **1967**, *43*, 287–360. [CrossRef]

27. Bale, J.; Hayward, S. Insect overwintering in a changing climate. *J. Exp. Biol.* **2010**, *213*, 980–994. [CrossRef]

28. Araj, S.E.; Wratten, S.; Lister, A.; Buckley, H.; Ghabeish, I. Searching behavior of an aphid parasitoid and its hyperparasitoid with and without floral nectar. *Biol. Control* **2011**, *57*, 79–84. [CrossRef]

29. Mitsumasu, K.; Azuma, M.; Niimi, T.; Yamashita, O.; Yaginuma, T. Membrane-penetrating trehalase from silkworm *Bombyx mori*: Molecular cloning and localization in larval midgut. *Insect Mol. Biol.* **2005**, *14*, 501–508. [CrossRef]

30. Santos, R.; Alves-Bezerra, M.; Rosas-Oliveira, R.; Majerowicz, D.; Meyer-Fernandes, J.R.; Gondim, K.C. Gene identification and enzymatic properties of a membrane bound trehalase from the ovary of *Rhodnius prolixus*. *Arch. Insect Biochem.* **2012**, *81*, 199–213. [CrossRef]

31. Shen, Q.D.; Yang, M.M.; Xie, G.Q.; Wang, H.J.; Zhang, L.; Qiu, L.Y.; Wang, S.G.; Tang, B. Excess trehalose and glucose affects chitin metabolism in brown planthopper (*Nilaparvata lugens*). *J. Asia Pac. Entomol.* **2017**, *20*, 449–455. [CrossRef]

32. Dindo, M.L.; Grenier, S. Production of Dipteran Parasitoids. In *Mass Production of Beneficial Organisms: Invertebrates and Entomopathogens*; Morales-Ramos, J.A., Guadalupe Rojas, M., Shapiro-Ilan, D.I., Eds.; Academic Press: London, UK, 2014; pp. 101–143.

33. Dindo, M.L.; Stangolini, L.; Marchetti, E. A simplified artificial medium for the In vitro rearing of *Exorista larvarum* (Diptera: Tachinidae). *Biocont. Sci. Tech.* **2010**, *20*, 407–410. [CrossRef]

34. Chen, Q.; Ma, E.; Behar, K.L.; Xu, T.; Haddad, G.G. Role of trehalose phosphate synthase in anoxia tolerance and development in *Drosophila melanogaster*. *J. Biol. Chem.* **2002**, *277*, 3274–3279. [CrossRef] [PubMed]

35. Chung, J.S. A trehalose 6-phosphate synthase gene of the hemocytes of the blue crab, *Callinectes sapidus*: Cloning, the expression, its enzyme activity and relationship to hemolymph trehalose levels. *Saline Syst.* **2008**, *4*, 18. [CrossRef]

36. Campbell, J.A.; Davies, G.J.; Bulone, V.; Henrissat, B.V. A classification of nucleotide-diphospho-sugar glycosyltransferases based on amino acid sequence similarities. *Biochem. J.* **1997**, *326*, 929–939. [CrossRef]

37. Zhu, K.Y.; Merzendorfer, H.; Zhang, W.Q.; Zhang, J.Z.; Muthukrishnan, S. Biosynthesis, turnover, and functions of chitin in insects. *Annu. Rev. Entomol.* **2006**, *61*, 177–196. [CrossRef]

38. Qin, Z.; Wang, S.; Wei, P.; Xu, C.D.; Tang, B.; Zhang, F. Molecular cloning and expression in cold induction of trehalose-6-phosphate synthase gene in *Harmonia axyridis* (Pallas). *Acta Entomol. Sin.* **2012**, *55*, 651–658. [CrossRef]

39. Xu, J.; Bao, B.; Zhang, Z.F.; Yi, Y.Z.; Xu, W.H. Identification of a novel gene encoding the trehalose phosphate synthase in the cotton bollworm, *Helicoverpa armigera*. *Glycobiology* **2009**, *19*, 250–257. [CrossRef] [PubMed]

40. Tang, B.; Chen, J.; Yao, Q.; Pan, Z.Q.; Xu, W.H.; Wang, S.G.; Zhang, W.Q. Characterization of a trehalose-6-phosphate synthase gene from *Spodoptera exigua* and its function identification through RNA interference. *J. Insect Physiol.* **2010**, *56*, 813–821. [CrossRef]

41. Tang, B.; Zheng, H.Z.; Xu, Q.; Zou, Q.; Wang, G.J.; Zhang, F.; Wang, S.G.; Zhang, Z.H. Cloning and pattern of expression of trehalose-6-phosphate synthase cDNA from *Catantops pinguis* (Orthoptera: Catantopidae). *Eur. J. Entomol.* **2011**, *108*, 355–363. [CrossRef]

42. Kern, C.; Wolf, C.; Bender, F.; Berger, M.; Noack, S.; Schmalz, S.; Ilg, T. Trehalose-6-phosphate synthase from the cat flea *Ctenocephalides felis* and *Drosophila melanogaster*: Gene identification, cloning, heterologous functional expression and identification of inhibitors by high throughput screening. *Insect Mol. Biol.* **2012**, *21*, 456–471. [CrossRef]

43. Tang, B.; Chen, X.; Liu, Y.; Tian, H.; Liu, J.; Hu, J.; Xu, W.; Zhang, W. Characterization and expression patterns of a membrane-bound trehalase from *Spodoptera exigua*. *BMC Mol. Biol.* **2008**, *9*, 51. [CrossRef] [PubMed]

44. Tatun, N.; Singtripop, T.; Tungjitwitayakul, J.; Sakurai, S. Regulation of soluble and membrane-bound trehalase activity and expression of the enzyme in the larval midgut of the bamboo borer *Omphisa fuscidentalis*. *Insect Biochem. Mol. Biol.* **2008**, *38*, 788–795. [CrossRef]

45. Gu, J.; Shao, Y.; Zhang, C.; Liu, Z.; Zhang, Y. Characterization of putative soluble and membrane-bound trehalases in a hemipteran insect, *Nilaparvata lugens*. *J. Insect Physiol.* **2009**, *55*, 997–1002. [CrossRef]

46. Tang, B.; Wei, P.; Chen, J.; Wang, S.G.; Zhang, W.Q. Progress in gene features and functions of insect trehalases. *Acta Entomol. Sin.* **2012**, *55*, 1315–1321. [CrossRef]

47. Parkinson, N.M.; Conyers, C.M.; Keen, J.N.; MacNicoll, A.D.; Smith, I.; Weaver, R.J. cDNAs encoding large venom proteins from the parasitoid wasp *Pimpla hypochondriaca* identified by random sequence analysis. Comp. Biochem. *Physiol. C Toxicol. Pharmacol.* **2003**, *134*, 513–520. [CrossRef]

48. Kumar, S.; Stecher, G.; Tamura, K. MEGA7: Molecular Evolutionary Genetics Analysis Version 7.0 for bigger datasets. *Mol. Biol. Evol.* **2016**, *33*, 1870–1874. [CrossRef]

49. Higgins, D.G.; Bleasby, A.J.; Fuchs, R. CLUSTAL V: Improved software for multiple sequence alignment. *Comput. Appl. Biosci.* **1992**, *8*, 189–191. [CrossRef]

50. Julie, D.T.; Arnaud, M.; Andrew, W.; Jim, P.; Geoffrey, J.B.; Frédéric, P.; Olivier, P. MACSIMS: Multiple alignment of complete sequences information management system. *BMC Bioinform.* **2006**, *2*, 317–318. [CrossRef]

51. Naruya, S.; Masatoshi, N. The neighbor-joining method: A new method for reconstructing phylogenetic trees. *Mol. Biol. Evol.* **1987**, *4*, 406–425. [CrossRef]

52. Hou, Y.B.; Zhang, H.; Miranda, L.; Lin, S.J. Serious overestimation in quantitative PCR by circular (Supercoiled) plasmid standard: Microalgal pcna as the model gene. *PLoS ONE* **2010**, *5*, e9545. [CrossRef]

53. Godornes, C.; Leader, B.T.; Molini, B.J.; Centurion-Lara, A.; Lukehart, S.A. Quantitation of rabbit cytokine mRNA by Real-Time RT-PCR. *Cytokine* **2007**, *38*, 1–7. [CrossRef]

54. Duan, F.; Gao, Y.H.; Yuan, J.G. Research on activity measurement of trehalose synthase. *Food Drug* **2008**, *10*, 47–50. [CrossRef]

55. Miller, G.L. Use of dinitrosalicylic acid reagent for determination of reducing sugar. *Anal. Chem.* **1959**, *31*, 426. [CrossRef]

56. Wang, N.S. Experiment no. 4A: Glucose Assay by Dinitrosalicylic Colorimetric Method. 2004. Available online: http://www.eng.umd.edu/~{}nsw/ench485/lab4a.htm (accessed on 27 January 2015).

57. Su, Z.H.; Ikeda, M.; Sato, Y.; Saito, H.; Imai, K.; Isobe, M.; Yamashita, O. Molecular characterization of ovary trehalase of the silkworm, *Bombyx mori* and its transcriptional activation by diapause hormone. *Biochim. Biophys. Acta* **1994**, *1218*, 366–374. [CrossRef]

58. Becker, A.; Schlöder, P.; Steele, J.; Wegener, G. The regulation of trehalose metabolism in insects. *Experientia* **1996**, *52*, 433–439. [CrossRef]

59. Shi, Z.K.; Liu, X.J.; Xu, Q.Y.; Qin, Z.; Wang, S.; Zhang, F.; Wang, S.G.; Tang, B. Two novel soluble trehalase genes cloned from *Harmonia axyridis* and regulation of the enzyme in a rapid changing temperature. *Comp. Biochem. Physiol.* **2016**, *198*, 10–18. [CrossRef]

Publisher's Note: MDPI stays neutral with regard to jurisdictional claims in published maps and institutional affiliations.

insects

Article

Effect of Prey Species and Prey Densities on the Performance of Adult *Coenosia attenuata*

Deyu Zou [1], Thomas A. Coudron [2], Lisheng Zhang [3], Weihong Xu [1], Jingyang Xu [1], Mengqing Wang [3], Xuezhuang Xiao [4] and Huihui Wu [4],*

1   Biological Control of Insects Research Laboratory, Institute of Plant Protection,
    Tianjin Academy of Agricultural Sciences, Tianjin 300384, China; zdyqiuzhen@126.com (D.Z.);
    xwh310@163.com (W.X.); jingyang_x@163.com (J.X.)
2   Biological Control of Insects Research Laboratory, USDA–Agricultural Research Service,
    Columbia, MO 65203, USA; coudront@missouri.edu
3   Key Laboratory of Integrated Pest Management in Crops, Ministry of Agriculture,
    Institute of Plant Protection, Chinese Academy of Agricultural Sciences, Beijing 100193, China;
    zhanglisheng@caas.cn (L.Z.); wangmengqing@caas.cn (M.W.)
4   College of Horticulture and Landscape, Tianjin Agricultural University, Tianjin 300392, China;
    xxz24250@163.com
*   Correspondence: wuhuihui@tjau.edu.com; Tel.: +86-22-23781319

Citation: Zou, D.; Coudron, T.A.;
Zhang, L.; Xu, W.; Xu, J.; Wang, M.;
Xiao, X.; Wu, H. Effect of Prey Species
and Prey Densities on the
Performance of Adult *Coenosia
attenuata*. *Insects* **2021**, *12*, 669.
https://doi.org/10.3390/
insects12080669

Academic Editor: Allen
Carson Cohen

Received: 31 May 2021
Accepted: 21 July 2021
Published: 23 July 2021

Publisher's Note: MDPI stays neutral
with regard to jurisdictional claims in
published maps and institutional affiliations.

Simple Summary: The predaceous fly *Coenosia attenuata* Stein has received attention because of its ability to effectively suppress a wide range of agricultural pests, such as fungus gnats, whiteflies and leaf miners. An effective level of control requires large numbers of *C. attenuata* to be available at low cost for release. Adult fungus gnats and drosophilids are now the main prey used to rear *C. attenuata* adults. However, previous studies showed *C. attenuata* fertility is lower when fed drosophilids compared to fungus gnats. The current study investigated the performance of *C. attenuata* adults when reared on different densities of adult *Drosophila melanogaster* Meigen or *Bradysia impatiens* (Johannsem). Results showed that the optimal prey density in the mass rearing of adult *C. attenuata* was 12–24 adult *B. impatiens* daily per predator. Additionally, *C. attenuata* adults suffered more wing damage, at some of the prey densities, when reared on *D. melanogaster* compared to *B. impatiens*. This information will be used to optimize rearing methods and decrease the cost of mass rearing in *C. attenuata*.

Abstract: Mass production of *Coenosia attenuata* Stein at low cost is very important for their use as a biological control agent. The present study reports the performance of *C. attenuata* adults when reared on *Drosophila melanogaster* Meigen or *Bradysia impatiens* (Johannsem). Different densities (6, 9, 15, 24 and 36 adults per predator) of *D. melanogaster* or (6, 12, 24, 36 and 48 adults per predator) of *B. impatiens* were used at 26 ± 1 °C, 14:10 (L:D) and 70 ± 5% RH. The results concluded that *C. attenuata* adults had higher fecundity, longer longevity and less wing damage when reared on *B. impatiens* adults compared to *D. melanogaster* adults. Additionally, *C. attenuata* adults demonstrated greater difficulty catching and carrying heavier *D. melanogaster* adults than lighter *B. impatiens* adults. In this case, 12 to 24 adults of *B. impatiens* daily per predator were considered optimal prey density in the mass rearing of adult *C. attenuata*.

Keywords: *Coenosia attenuata*; mass rearing; wing damage; *Bradysia impatiens*; *Drosophila melanogaster*; fecundity

1. Introduction

The predaceous fly *Coenosia attenuata* Stein (Diptera: Muscidae), also known as "tiger fly", "killer fly" or "hunter fly" [1–4], is native to Southern Europe [4,5] and has been reported to have spontaneously colonized a number of crops outdoors and in greenhouses in many countries worldwide [5–21]. It has received attention because of its ability to effectively suppress a wide range of agricultural pests, such as fungus gnats (Diptera: Sciaridae), whiteflies (Hemiptera: Aleyrodidae), leaf miners (Diptera: Agromyzidae), winged aphids (Hemiptera: Aphididae), leafhoppers of the genera *Eupteryx* (Hemiptera: Cicadellidae) and *Empoasca* (Hemiptera: Cicadellidae), midges (Diptera: Chironomidae), moth flies (Diptera: Psychodidae), shore flies (Diptera: Ephydridae) and fruit flies (Diptera: Drosophilidae) [7,8,13,22–35]. The wide range of prey used as food make the tiger-fly an attractive alternative to conventional control methods.

Intact wings play an important role in the life of *C. attenuata* adults. Adults of *C. attenuata* catch their prey while in flight and pursue targets at the range of 23–212 mm. Hence, they employ an interception strategy that is more energy efficient to intercept targets, which allows *C. attenuata* to cope with the extremely high line-of-sight rotation rates and thus prevents overcompensation of steering [36]. Adults of *C. attenuata* use mean flight speeds of 0.69 ms^{-1}, mean wingbeat frequency of 306, 19 Hz and acceleration of mean peak 9.3 ms^{-2} to intercept prey [36]. The flight of *C. attenuata* individuals is affected by environmental factors, adjusting in response to changes in temperature, the number of prey flights and conspecific density [37]. Therefore, wing damage will cause negative effects on the life of *C. attenuata* adults.

Mass rearing of *C. attenuata* is important given the environmental, health and resistance issues associated with the use of chemical insecticides. To achieve an effective level of control, however, requires the production of a large number of *C. attenuata* at low cost. Adults of fungus gnat and drosophilid are now the main prey used to rear *C. attenuata* adults [17,23,38,39]. Rearing drosophilids is quick, easy and not particularly expensive. However, they were used primarily as a complement to the fungus gnat diet because *C. attenuata* fertility is lower when fed drosophilids compared to fungus gnats [23]. The reason why the performance of *C. attenuata* reared on drosophilids is lower than those reared on fungus gnats have not been assessed. The present study reports our finding that *C. attenuata* adults had less wing damage, higher fecundity and longer longevity when reared on *Bradysia impatiens* (Johannsem) (Diptera: Sciaridae) compared to *Drosophila melanogaster* Meigen (Diptera: Drosophilidae).

2. Materials and Methods

2.1. Vinegar Flies

The vinegar fly, *D. melanogaster* were reared on bananas in open plastic canisters (about 1200 cm^3) in tissue bags (40 × 30 cm^2, 0.4-mm mesh openings) closed with binder clips. Adults were introduced into the tissue bags and the adults of the following generation started to emerge after ca. 11 days. The colony was maintained in a laboratory incubator and held at 26 ± 1 °C, 14:10 (L:D) and 70 ± 5% RH.

2.2. Fungus Gnats

A colony of fungus gnats was initiated with about 400 *B. impatiens* adults captured from a greenhouse at Wuqing Experiment Station (Tianjin, China). Fungus gnats were reared using the method very similar to that reported by Zou et al. (2021) [39]. Modifications were made to simplify and improve the processes of rearing and collecting fungus gnats for use in bioassays. Briefly, 300 mL of black peat (Lvdimeijing Science and Technology Co., Ltd., Beijing, China) and 55 to 60 g of dry kidney bean powder were placed in an open plastic box (25.5 × 19 × 7.8 cm^3). The mix was then moistened with 250 mL of tap water and 0.2-cm thick layer of moist coir (Shanghai Galuku Agricultural Science and Technology Co., Ltd. Shanghai, China; desalted, EC = 0.5, family pack, common grade) was placed on the top of the mix. Then the open plastic box was placed in a tissue bag (50 × 35 cm^2, 0.4-mm mesh openings). In this case, 400 to 500 newly emerged adult fungus gnats were placed in the tissue bag and closed with a binder clip. Fresh rearing medium was prepared daily and new cultures were set up daily.

The new fungus gnat adults deposited eggs on the media consisted of black peat, tap water and kidney bean powder. Newly hatched larvae fed on the media and adults emerged after 18–22 d. The colony was maintained in a laboratory incubator and held at 26 ± 1 °C, 14:10 (L:D) and 70 ± 5% RH.

2.3. Tiger-Fly

The *C. attenuata* used to establish a laboratory colony in this study were collected at Leizhuangzi flower farm of Tianjin, China. Adults were provided an oviposition tissue cage (60 × 55 × 50 cm^3), in which an open plastic box (29 × 20 × 7 cm^3) containing black peat, tap water, kidney bean powder and eggs of *B. impatiens*. A 0.5-cm thick layer of moist coir was placed on the top of the rearing media and used for oviposition. *B. impatiens* and *D. melanogaster* adults were supplied as prey daily. Five to 6 days later, the plastic boxes containing eggs of *C. attenuata* and rearing media were removed to another cage and second and third instar larvae of *B. impatiens* were added to the box to feed larvae of *C. attenuata*. Distilled water was added to the box when the media became dry. About 20 to 21 days later, adults of *C. attenuata* emerged. The colony was maintained in an artificial climate chamber and held at 26 ± 1 °C, 14:10 (L:D) and 70 ± 5% RH.

2.4. Performance of Tiger-Fly Adults Reared on Different Prey Species at Different Prey Densities

Five female/male pairs of newly emerged adults of *C. attenuata* (<24-h-old), in the first generation that originated from field-caught adults, were transferred into tissue cages (60 × 55 × 50 cm^3) containing an open 90-mm-diameter Petri dish containing of 0.7-cm thick layer of moist coir for oviposition. The moist coir was replenished after collecting egg daily. Two 110 cm strings were hung inside each cage and served as a perch for adult predators. In this case, 6, 9, 15, 24 and 36 adults of *D. melanogaster* and 6, 12, 24, 36 and 48 adults of *B. impatiens* (<24-h-old) were provided daily per predator adult. Prey were used only once and fresh prey were added daily. Five female/male pairs of *C. attenuata* adults (in one cage) were tested until death in each treatment and replicated 9 times. In total, 45 pairs of *C. attenuata* adults were tested in each treatment within a phytotron held at 26 ± 1 °C, 14:10 (L:D) and 70 ± 5% RH. Wing damage in *C. attenuata* adults was measured daily using a Mitutoyo 500-196-30 digital caliper (Mitutoyo, Kawasaki, Japan). Wing damage occurred along the long axis of wing. The extent of damage was calculated as a percentage using the length of damaged wing/the total length of wing × 100%. Both wings were assessed and the average was taken. The numbers of surviving and killed prey were recorded daily. Eggs deposited in the moist coir were collected and counted from each cage daily using a 00-sized paintbrush. They were then placed in 60-mm-diameter Petri dishes containing a single 55-mm-diameter filter paper, moistened with distilled water, sealed with Parafilm and inverted to keep the eggs moist. Hatch occurred ca. 6 days after oviposition and egg viability was calculated. Distilled water was sprayed to each tissue cage two times in the morning and afternoon per day. Tiger-flies were maintained in this manner until death.

2.5. Comparison of Body Weight and Body Length in C. attenuata, D. Melanogaster and B. Impatiens

The adults of *C. attenuata* caught prey in flight. So carrying prey with different body weights may cause different levels of wing damage for adults of *C. attenuata*. Adults of *C. attenuata* (<24-h-old) and *D. melanogaster* (<24-h-old) were placed in tissue bags (40 × 30 cm^2) and held in a Siemens BCD-501W fridge (Siemens, Nanjing, China) at −20 °C for 2 min before taking body length and weight measurements. Adults of *B. impatiens* (<24-h-old) were handled in the same way with *C. attenuata* and *D. melanogaster* and held at −20 °C for 3 min before using. Adult body length was measured (in resting position) from the apex of the head to the wing tip for *C. attenuata* and *D. melanogaster*. For *B. impatiens*, body length was measured (in resting position) from the apex of the head to the abdomen tip. Body length measurements were made using a Mitutoyo 500-196-30 digital caliper (Mitutoyo, Kawasaki, Japan). Weight measurements were made using a Sartorius BP 211D (Sartorius AG, Göttingen, Germany) balance. In this case, 30 females and 30 males were measured for each species.

2.6. Statistical Analyses

One-way ANOVA with subsequent Tukey's HSD test at $\alpha = 0.05$ was used to compare the proportion of damaged wing, number of prey killed, preovipositional period, total fecundity between different prey densities, body weight and body length between insect species. To avoid possible mistakes due to multiple testing of the same data base, the *p*-values were Bonferroni corrected. Two sample t-tests for means were used to compare proportion of damaged wing, preovipositional period and total fecundity between prey species. These comparisons were carried out on day 4 and the last day. The proportion of eggs successfully hatched was compared between treatments by the Chi-square test at $\alpha = 0.01$. All the statistical tests were carried out using SAS version 9.4.

3. Results

3.1. Comparison of Wing Damage of C. attenuata When Reared on Different Prey at Different Prey Densities

The mean proportion of damaged wings of *C. attenuata* females fed on *D. melanogaster* and males fed on *B. impatiens* or *D. melanogaster* did not differ significantly between prey densities at early ages (day 4) ($F_{4, 220} = 1.27$, $p = 0.2826$; $F_{4, 220} = 1.55$, $p = 0.1886$; $F_{4, 220} = 2.4$, $p = 0.0513$, respectively) (Figure 1B–D). However, *C. attenuata* females fed 48 adults of *B. impatiens* daily lost more wings compared with those fed 12 adults of *B. impatiens* daily on day 4 ($F_{1, 88} = 9.39$, $p = 0.0029$, Bonferroni-corrected $p = 0.005$) (Figure 1A). The wing damage increased with age in every case, but at different rates depending on prey species and density. When fed with *B. impatiens*, *C. attenuata* females showed significant differences at late age (31 days) between prey densities of 12 and 24 ($F_{1, 88} = 13.15$, $p = 0.0004$, Bonferroni-corrected $p = 0.005$), or between prey densities of 36 and 48 ($F_{1, 88} = 10.16$, $p = 0.002$, Bonferroni-corrected $p = 0.005$) (Figure 1A). For *C. attenuata* males, there were no significant differences between prey densities at late age (19 days) ($F_{4, 220} = 0.63$, $p = 0.6425$) (Figure 1B). When fed with *D. melanogaster*, both females and males of *C. coenosia* did not show significantly different wing damage between prey densities at later age (8 days) ($F_{4, 220} = 2.08$, $p = 0.0848$; $F_{4, 220} = 2.07$, $p = 0.0856$, respectively) (Figure 1C,D).

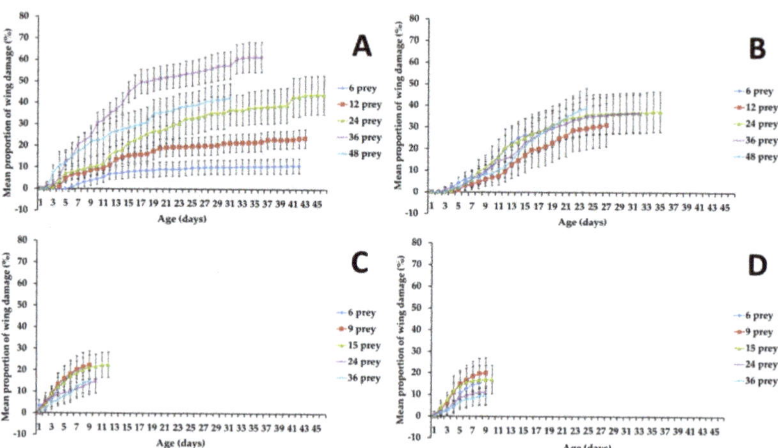

Figure 1. Mean proportion of damaged wings of all *Coenosia attenuata* when fed different prey at different densities daily per predator adult (mean values with 95% confidence intervals; error bars: 95% CI): (**A**) females fed 6, 12, 24, 36 and 48 adults of *Bradysia impatiens*; (**B**) males fed 6, 12, 24, 36 and 48 adults of *Bradysia impatiens*; (**C**) females fed 6, 9, 15, 24 and 36 adults of *Drosophila melanogaster*; (**D**) males fed 6, 9, 15, 24 and 36 adults of *Drosophila melanogaster*.

Females and males of *C. attenuata* had much shorter longevity when fed *D. melanogaster* adults compared to those fed *B. impatiens* adults (maximum age of 12 for *D. melanogaster* and of 46 for *B. impatiens* in females; maximum age of 10 for *D. melanogaster* and of 35 for *B. impatiens* in males). Mean proportion of wing damage in flies fed with *B. impatiens* at age 8 ranged from 3.25 to 22.42% in females and from 4.68 to 9.58% in males, while in flies fed with *D. melanogaster* it ranged from 12.28 to 21.56% in females and from 9.16 to 19.77% in males (Figure 1).

C. attenuata females fed 6 adults of *D. melanogaster* daily lost significantly more wings than those fed 6 adults of *B. impatiens* daily on day 4 ($t = -2.935$, $df = 88$, $p = 0.0043$) (Figure 1A,C). However, there was no significant difference in wing damage for *C. attenuata* females fed 24 adults of *D. melanogaster* compared with those fed 24 adults of *B. impatiens* daily, or for *C. attenuata* females fed 36 adults of *D. melanogaster* compared with those fed 36 adults of *B. impatiens* daily on day 4 ($t = -1.934$, $df = 88$, $p = 0.0564$; $t = -0.447$, $df = 88$, $p = 0.6561$, respectively) (Figure 1A,C). *C. attenuata* females lost significantly more wings when fed adults of *D. melanogaster* daily compared to *B. impatiens* at the prey density of 6 and 36 on day 8 ($t = -4.556$, $df = 88$, $p < 0.0001$; $t = 2.803$, $df = 88$, $p = 0.0062$, respectively). However, there was no significant difference in wing damage for *C. attenuata* females when fed 24 adults of *D. melanogaster* compared with those fed 24 adults of *B. impatiens* on day 8 ($t = -0.874$, $df = 88$, $p = 0.3847$) (Figure 1A,C).

There was no significant difference in wing damage for *C. attenuata* males fed 6 adults of *D. melanogaster* compared with those fed 6 adults of *B. impatiens*, or for *C. attenuata* males fed 36 adults of *D. melanogaster* compared with those fed 36 adults of *B. impatiens* on day 4 ($t = -1.35$, $df = 88$, $p = 0.1805$; $t = -1.1$, $df = 88$, $p = 0.2743$, respectively) (Figure 1B,D). *C. attenuata* males lost significantly more wings when fed adults of *D. melanogaster* daily compared to *B. impatiens* at the prey density of 24 on day 4 ($t = -2.302$, $df = 88$, $p = 0.0237$) and the prey density of 6 on day 8 ($t = -2.631$, $df = 88$, $p = 0.01$). However, there was no significant difference in wing damage for *C. attenuata* males when fed adults of *D. melanogaster* daily compared to *B. impatiens* at the prey density of 24 or 36 on day 8 ($t = -0.362$, $df = 88$, $p = 0.7181$; $t = -0.482$, $df = 88$, $p = 0.6311$, respectively) (Figure 1B,D).

3.2. Number of Prey killed by C. attenuata Reared on Different Prey at Different Prey Densities

C. attenuata adults killed all *B. impatiens* adults when fed 6 prey daily per predator adult except for the last 3 days, which suggests 6 adults of *B. impatiens* are not enough for *C. attenuata* adults (Figure 2). Most of *B. impatiens* adults were killed by *C. attenuata* adults when fed 12 prey daily per predator adult. The number of prey killed by *C. attenuata* adults fluctuated when fed 24, 36 and 48 prey daily per predator adult. The numbers of prey killed per predator daily were 15.33 to 23.77, 25.99 to 35.70 and 33.00 to 47.45 for *C. attenuata* adults fed 24, 36 and 48 prey daily per predator adult, respectively. The number of prey killed daily per predator decreased in the last few days with the increase of age and the mean proportion of damaged wings for *C. attenuata* adults fed 12, 24, 36 and 48 prey daily per predator adult (Figure 2). The number of *B. impatiens* adults killed daily per predator adult differed significantly between prey densities at early ages (day 4) ($p < 0.0001$ for 9 comparisons), except for densities of 24 vs. 36 ($F_{1,\,16} = 5.61$, $p = 0.0308$, Bonferroni-corrected $p = 0.005$) (Figure 2). At later age (day 31), the number of *B. impatiens* adults killed daily per predator adult differed significantly between prey densities ($p < 0.0001$ for 9 comparisons), except for densities of 36 vs. 48 ($F_{1,\,4} = 0.26$, $p = 0.6392$, Bonferroni-corrected $p = 0.005$) (Figure 2).

C. attenuata adults killed 3.17 to 4.52 *D. melanogaster* adults when fed 6 prey adults daily per predator adult (Figure 3). The number of prey killed by *C. attenuata* adults fluctuated when fed 9, 15, 24 and 36 prey daily per predator adult. The numbers of prey killed daily per predator were 4.27 to 6.54, 4.37 to 7.29, 7.57 to 11.98 and 9.58 to 15.55 for *C. attenuata* adults fed 9, 15, 24 and 36 prey daily per predator adult, respectively. The number of prey killed per predator daily decreased in the last few days with the increase of age and the proportion of broken wings for *C. attenuata* adults in all treatments. The number of *D. melanogaster* adults killed daily per predator adult differed significantly with prey densities of 6 vs. 9 ($F_{1,\,16} = 10.64$, $p = 0.0049$, Bonferroni-corrected $p = 0.005$), 9 vs. 36 ($F_{1,\,16} = 17.21$, $p = 0.0008$,

Bonferroni-corrected $p = 0.005$) and with the prey densities of 6 vs. 15, 6 vs. 24, 6 vs. 36 and 15 vs. 36 (all $p < 0.0001$) at early ages (day 4). No significant differences were found in the other 4 comparisons at early ages (day 4) (Figure 3). Additionally, no significant differences were found in all 10 comparisons at later age (day 8) (Figure 3).

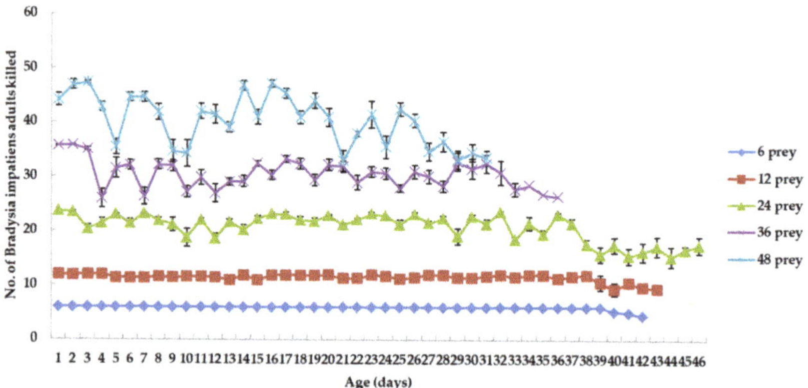

Figure 2. Mean number of prey killed per *C. attenuata* adult when fed 6, 12, 24, 36 and 48 adults of *B. impatiens* daily per predator adult (values are mean ± SE, single values were present when only one adult *C. attenuata* was left in the end).

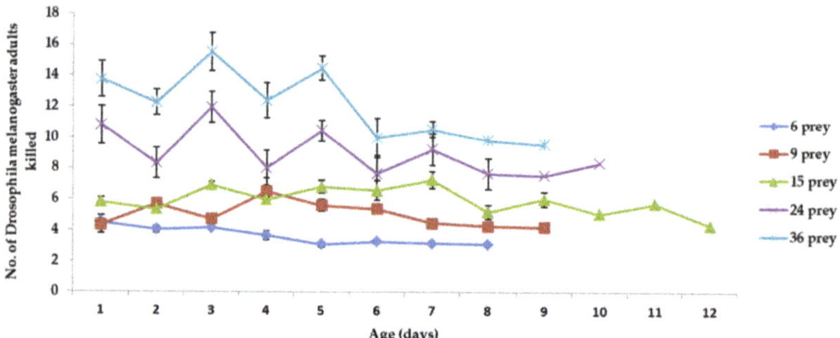

Figure 3. Mean number of prey killed per *C. attenuata* adult when fed 6, 9, 15, 24 and 36 adults of *D. melanogaster* daily per predator adult (values are mean ± SE, single values were present when only one adult *C. attenuata* was left in the end).

3.3. Preovipositional Period of C. attenuata Female Reared on Different Prey at Different Prey Densities

There were no significant differences in preovipositional period in any of the treatments of *C. attenuata* females when fed 6, 9, 15, 24 or 36 adults of *D. melanogaster* prey ($F_{4, 40} = 2.43$, $p = 0.064$) (Figure 4). The preovipositional period of *C. attenuata* female was from 4.22 to 5.22 days when fed *D. melanogaster* prey. Similar, there were no significant differences in preovipositional period in any of the treatments of *C. attenuata* females when fed 6, 12, 24, 36 or 48 adults of *B. impatiens* ($F_{4, 40} = 1.73$, $p = 0.162$). The preovipositional period of *C. attenuata* female was from 3.89 to 4.67 days when fed *B. impatiens* prey. There were no significant differences in the preovipositional period for *C. attenuata* females when fed 6 adults of *D. melanogaster* prey or 6 adults of *B. impatiens* prey, or for *C. attenuata* females when fed 36 adults of *D. melanogaster* prey or 36 adults of *B. impatiens* prey ($t = -0.686$, $df = 16$, $p = 0.5025$; $t = -1.715$, $df = 16$, $p = 0.1056$, respectively) (Figure 4). The preovipositional periods were significantly longer for *C. attenuata* females fed 24 adults of *D. melanogaster* prey than of *C. attenuata* females fed 24 adults of *B. impatiens* prey. These differences, although statistically significant, were small ($t = -2.132$, $df = 16$, $p = 0.0489$) (Figure 4).

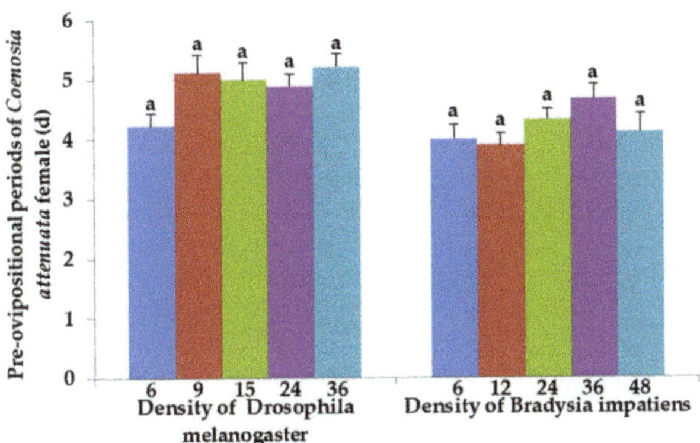

Figure 4. Preovipositional periods of *C. attenuata* female adult when fed 6, 9, 15, 24 and 36 adults of *D. melanogaster* and 6, 12, 24, 36 and 48 adults of *B. impatiens* daily per predator adult. Different letters above each bar indicate significant differences between prey densities using one-way ANOVA, Tukey's HSD test ($p = 0.05$ and n = 9).

3.4. Total Fecundity of C. attenuata Female Reared on Different Prey at Different Prey Densities

The total fecundity per female of *C. attenuata* fed *D. melanogaster* prey differed significantly with prey densities of 6 vs. 36 ($F_{1, 16} = 17.38$, $p = 0.0007$, Bonferroni-corrected $p = 0.005$), 9 vs. 36 ($F_{1, 16} = 19.82$, $p = 0.0004$, Bonferroni-corrected $p = 0.005$), 24 vs. 36 ($F_{1, 16} = 18.90$, $p = 0.0005$, Bonferroni-corrected $p = 0.005$) and the other 4 comparisons (6 vs. 9, 6 vs. 15, 6 vs. 24 and 15 vs. 36, all $p < 0.0001$). No significant differences were found in other 3 comparisons (Figure 5). The total fecundity per female of *C. attenuata* fed *B. impatiens* prey did not differ significantly with prey densities of 6 vs. 48 and 12 vs. 36 ($F_{1, 16} = 0.17$, $p = 0.6829$; $F_{1, 16} = 9.54$, $p = 0.007$, respectively, Bonferroni-corrected $p = 0.005$). However, the total fecundity per female of *C. attenuata* fed *B. impatiens* prey differed significantly between prey densities for the other 8 comparisons (all $p < 0.0001$) (Figure 5). Additionally, the total fecundity was much higher for females of *C. attenuata* fed *B. impatiens* adults than for those fed *D. melanogaster* prey at the same prey densities of 6, 24 and 36 ($t = 31.971$, $df = 16$, $p < 0.0001$; $t = 36.609$, $df = 16$, $p < 0.0001$; $t = 32.954$, $df = 16$, $p < 0.0001$, respectively).

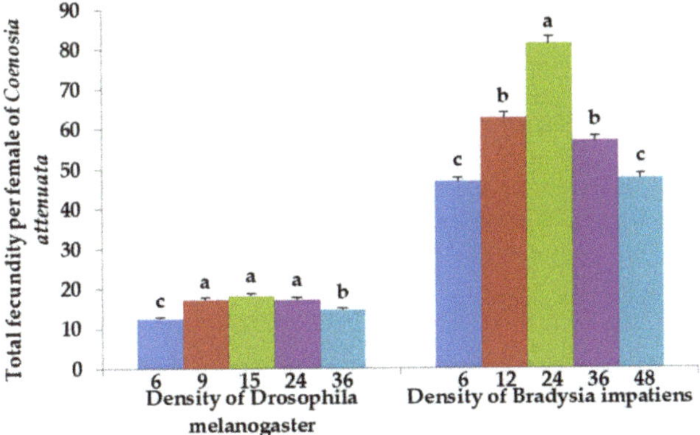

Figure 5. Total fecundity per female of *C. attenuata* when fed 6, 9, 15, 24 and 36 adults of *D. melanogaster* and 6, 12, 24, 36 and 48 adults of *B. impatiens* daily per predator adult. Different letters above each bar indicate significant differences between prey densities (corrected p value for multiple testing by Bonferroni correction is 0.005 and n = 9).

3.5. Proportion of Eggs Successfully Hatched in C. attenuata Reared on Different Prey at Different Prey Densities

There was a significantly higher proportion of eggs that successfully hatched for *C. attenuata* fed 9 adults of *D. melanogaster* than for those fed 24 and 36 adults of *D. melanogaster* daily per predator adult ($\chi^2 = 8.37$, $p = 0.004$; $\chi^2 = 20.56$, $p < 0.0001$, respectively) (Figure 6). However, there were no significant differences in the proportion of eggs that successfully hatched between *C. attenuata* fed 9 adults of *D. melanogaster* and those fed 6 or 15 adults of *D. melanogaster* ($\chi^2 = 6.36$, $p = 0.012$; $\chi^2 = 0.81$, $p = 0.368$, respectively). There were no significant differences in proportion of eggs successfully hatched between *C. attenuata* fed 24 adults of *B. impatiens* and those fed 6, 12 or 36 adults of *B. impatiens* ($\chi^2 = 5.97$, $p = 0.015$; $\chi^2 = 1.28$, $p = 0.257$; $\chi^2 = 2.92$, $p = 0.087$, respectively). The proportion of eggs that successfully hatched was significantly higher for *C. attenuata* fed 24 adults of *B. impatiens* than for those fed 48 adults of *B. impatiens* daily per predator adult ($\chi^2 = 6.87$, $p = 0.009$). Additionally, the proportion of eggs that successfully hatched was much higher for *C. attenuata* adults fed *B. impatiens* adults than for those fed *D. melanogaster* adults at the same prey densities of 6, 24 and 36 ($\chi^2 = 15.14$, $p < 0.0001$; $\chi^2 = 43.35$, $p < 0.0001$; $\chi^2 = 43.26$, $p < 0.0001$, respectively).

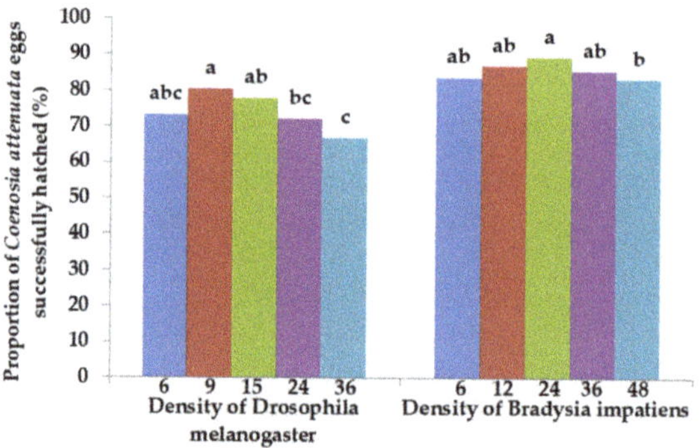

Figure 6. Proportion of eggs successfully hatched for *C. attenuata* when fed 6, 9, 15, 24 and 36 adults of *D. melanogaster* and 6, 12, 24, 36 and 48 adults of *B. impatiens* daily per predator adult. Different letters above each bar indicate significant differences between prey densities at a 0.01 level of significance using Chi-square test (n = 450, first 50 eggs were collected from each cage, 9 repetitions in each treatment).

3.6. Comparation of Body Weight and Body Length in C. attenuata, D. melanogaster and B. impatiens

Adult females of *C. attenuata* were significantly longer than those of *D. melanogaster* and *B. impatiens* ($F_{2, 87} = 1335.80$, $p < 0.0001$) (Table 1). Adult females of *D. melanogaster* were significantly longer than those of *B. impatiens* ($F_{2, 87} = 1335.80$, $p < 0.0001$). Similar, adult males of *C. attenuata* were significantly longer than those of *D. melanogaster* and *B. impatiens* ($F_{2, 87} = 2101.27$, $p < 0.0001$) and the body length was significantly longer for adult males of *D. melanogaster* than for adult males of *B. impatiens* ($F_{2, 87} = 2101.27$, $p < 0.0001$).

Table 1. Body length (mm) and body weight (mg) of adult *Coenosia attenuata*, *Drosophila melanogaster* and *Bradysia impatiens* (n = 30).

Insects	Body Length of Adult Female [a] (mm)	Body Length of Adult Male [a] (mm)	Body Weight of Adult Female [b] (mg)	Body Weight of Adult Male [b] (mg)
Coenosia attenuata	4.49 ± 0.04 [a]	3.68 ± 0.02 [a]	2.52 ± 0.04 [a]	1.47 ± 0.03 [a]
Drosophila melanogaster	3.12 ± 0.02 [b]	2.78 ± 0.02 [b]	1.09 ± 0.01 [b]	0.72 ± 0.01 [b]
Bradysia impatiens	2.22 ± 0.02 [c]	1.70 ± 0.02 [c]	0.42 ± 0.01 [c]	0.23 ± 0.01 [c]

Values are mean ± SE. Means in columns with the same letter are not significantly different at a 0.05 level of significant. [a] Adult body length measured < 24 h after adult emergence. [b] Adult body weight measured < 24 h after adult emergence.

Adult females of *C. attenuata* were significantly heavier than those of *D. melanogaster* and *B. impatiens* and there was a significant difference in the body weight of adult females between *D. melanogaster* and *B. impatiens* ($F_{2, 87}$ = 2188.57, p < 0.0001). Similar, adult males of *C. attenuata* were significantly heavier than those of *D. melanogaster* and *B. impatiens* and there was a significant difference in the body weight of adult males between *D. melanogaster* and *B. impatiens* ($F_{2, 87}$ = 1164.34, p < 0.0001) (Table 1).

4. Discussion

The flight of *C. attenuata* individuals was affected by environmental factors and was increased in response to increases in the number of prey flights [37]. Bonsignore (2016) found that predatory flights of adult *C. attenuata* comprised a small percentage (ca. 6%) of the total flights, with a predation success rate of 61% [37]. In our study, the mean proportion of damaged wings of *C. attenuata* females when fed 6, 12, 24 and 36 adults of *B. impatiens* daily per predator adult was increased in response to increases in the number of prey densities. However, the mean proportion of wing damage in *C. attenuata* females was lower for prey densities of 48 adults of *B. impatiens* than for prey densities of 36. The high density of 48 adults of *B. impatiens* probably increased the predation success rate and thereby decreased the mean proportion of damaged wings of *C. attenuata* female although the tiger-fly is regarded to have predation instinct [40,41]. Damaged wings of *C. attenuata* males fed on *B. impatiens* continued to increase with an increase in age of *C. attenuata*. However, the mean proportions of damaged wings in *C. attenuata* males were not consistent with those of females. Prey density did not cause significant effect on wing damage for *C. attenuata* males, which suggests prey density was not the only factor affecting wing damage in *C. attenuata* males. Being attacked by female *C. attenuata* and attempting to mate with female *C. attenuata* could also influence wing damage. Additionally, male adults required less prey compared to female adults, which means low prey densities could increase the predation success rate and thereby decreased the proportion of damaged wings of *C. attenuata* males. According to the damaged wings and longevity of *C. attenuata* adults, prey densities of 12 to 24 should be optimal density for mass rearing of adult *C. attenuata*.

Prey density of vinegar fly did not cause a significant effect on the mean proportion of damaged wings in both female and male adults of *C. attenuata*. It seems reasonable to conclude that the short lifespan of the tiger-fly was too short to manifest an effect of *D. melanogaster* prey density on wing damage of *C. attenuata* adults. *C. attenuata* adults fed *D. melanogaster* prey daily lost more wings compared to those fed *B. impatiens* prey at the same age for some prey density. This may be related to the increased difficulty in carrying heavier *D. melanogaster* adults than *B. impatiens* adults.

Bonsignore (2016) sorted adult *C. attenuata* flights into three groups, movement flights, territory defense flights and predatory flights in greenhouse [37]. However, we observed that there should be another type of flight, escape flights in cage. We hypothesize that adult *C. attenuata* want to escape from the cage when encountering high density of prey, resulting in less wing damage. Escaping from the environment with high prey density may be a self-protection response for adult *C. attenuata*.

We found that females lived longer than males, as reported by Kühne et al. (1997) [23]. However, these authors record 38 days and 33 days as the maximum female and male longevity, respectively, under laboratory conditions (25 °C and 50–60% RH) and an estimated longevity of eight weeks under greenhouse conditions. Predators fed *B. impatiens* adults in our study lived 46 days and 35 days as the maximum female and male longevity, respectively, possibly because they had a better food supply. However, in our study, predators fed adult *D. melanogaster* flies lived only 12 days and 10 days as the maximum female and male longevity, respectively. We speculate that *C. attenuata* adults were able to more easily capture lighter *B. impatiens* adults than heavier adult *D. melanogaster*. Additionally, adult *C. attenuata* were able to attack adult *B. impatiens* on the bottom of cage when they could only jump or crawl because of damaged wings. However, it is difficult for *C. attenuata* with damaged wings to capture adult *D. melanogaster*.

Female adult *C. attenuata* were found to exhibit a type I functional response to adult sciarid flies, which was conducted in glass vials 8 cm long and 8 cm in diameter at 25 °C at 60–80% RH, with a 16L:8D photoperiod. Sciarids were consumed in significantly different numbers at densities from 5 to 20 individuals (the number of killed flies changed from 2.90 to 8.4, respectively). However, increasing prey availability beyond 20 individuals resulted in no substantial increase in predation [30]. However, female adult *C. attenuata* were found to exhibit a type II functional response to adult *D. melanogaster* flies, which was conducted in Plexiglas cages with a dimension of 25 by 25 by 25 cm at 30 °C at 65 ± 5% RH, with a 12L:12D photoperiod. *D. melanogaster* flies were consumed in significantly different numbers at densities from 5 to 55 individuals (the number of attacked flies changed from 3.50 to 5.67, respectively) [42]. Kühne (2000) states that each adult *C. attenuata* needs either 1.5 adults of *D. melanogaster* or 6.9 adults of *B. impatiens* per day [25]. We did not analyze the functional response to adult *B. impatiens* or *D. melanogaster* flies because more than one factor affected functional response, such as intraspecific competition and predation. The number of killed prey in our study was more than those mentioned above, which was probably caused by intraspecific competition and predation instinct resulting from cage and space differences. The flight ability of adult *B. impatiens* is weak and often some of them stayed on the bottom of cage which made it more convenient for adult *C. attenuata* without flight ability, because of damaged wings, to catch the adult *B. impatiens*. In contrast, the flight ability of adult *D. melanogaster* is strong and it is more difficult for adult *C. attenuata* with weakened flight ability to catch adult *D. melanogaster*, although adult *C. attenuata* has been proved to be more efficient in information sampling and processing than adult *D. melanogaster* [43,44].

The preoviposition period of *C. attenuata* is approximately 4 days [23]. Our reports showed similar preoviposition periods when fed adults of *B. impatiens* with 3.89–4.67 days and adults of *D. melanogaster* with 4.22–5.22 days. Prey density, prey species and damaged wings did not cause negative effects on the preoviposition period of *C. attenuata*. Sanderson et al. (2009) found the tiger-flies laid more eggs with fungus gnat prey than shore fly prey [27]. We found that the tiger-flies laid much more eggs with fungus gnat prey than vinegar fly prey. However, the total fecundity per female of *C. attenuata* did not continue to increase with an increase in prey density. Shorter life span probably cause the lower fecundity for *C. attenuata* female when fed adult *D. melanogaster* compared to adult *B. impatiens*. Martins et al. (2015) presented an optimized method for mass rearing *C. attenuata* with fungus gnats and Drosophilids as prey, where the number of adults that emerged per parental pair ranged from 1.8 to 9.0 (= per pair progeny production, or the number of adult offspring that emerged in each cage divided by the number of parental pairs) [38]. In our study, the number of adult offspring that emerged ranged from 4.96 to 7.64 and 21.16 to 39.27 at least for per parental pair when fed adult *D. melanogaster* and *B. impatiens*, respectively according to survival rates of larvae, percentages of pupation and adult emergence in our previous reports [17,39]. The proportion of eggs that successfully hatched was much higher for *C. attenuata* adults fed *B. impatiens* adults than for those fed *D. melanogaster* adults at the same prey densities. Longer longevity in male *C. attenuata* and lighter body weight in *B. impatiens* prey correlated to increased proportion of eggs that successfully hatched in *C. attenuata*.

Predation by adult *C. attenuata* is rapid, and adults take off as soon as they observe their prey in flight, although they do not know the absolute size of the potential prey prior to the flight [45]. One important physical factor affecting predator responses is prey size [46]. Body length and body weight of adult *C. attenuata*, *D. melanogaster* and *B. impatiens* were analyzed in our study to better understand the complexity of predation. We report body length and body weight of adult *C. attenuata* from Tianjin to be similar to those reported by us previously [17,39] and to those reported in Uruguay where the *C. attenuata* flies measured approximately 2.5–5.00 mm in length [47]. Body weight of adult *D. melanogaster* measured in this study is similar to those reported by Chen et al. (2019) [48]. Body length of adult *B. impatiens* analyzed in this study is similar to those reported by Wilkinson and

Daugherty (1970) [49]. Obviously, it is more difficult for adult *C. attenuata* to catch and carry heavier adult *D. melanogaster* than lighter adult *B. impatiens*. Most importantly, it is more difficult for male adult *C. attenuata* to catch and carry female adult *D. melanogaster*, that are 74.15% weight of male adult of *C. attenuata* than to catch adult female *B. impatiens* that only weigh 28.57% of their weight.

In our study, we demonstrated that adult *B. impatiens* was an optimal prey in the mass rearing of adult *C. attenuata* although rearing drosophilids is quick, easy and not particularly expensive. In addition, we provide evidence that damage to wings of adult *C. attenuata* when fed adult *D. melanogaster* vs. *B. impatiens* is an important consideration for prey selection. We conclude a prey density of 12–24 adult fungus gnats daily per adult predator as optimal for mass rearing of adult *C. attenuata*. Rearing cost, nutritional difference, digestion efficiency, chemical, morphological and behavioral defense mechanisms of a prey will be explored in future studies.

5. Conclusions

We present the first report of wing damage for *C. attenuata* adults when reared on different prey. The results indicate that *C. attenuata* adults had higher fecundity, longer longevity and generally less wing damage when reared on *B. impatiens* compared to *D. melanogaster*. Lighter body weight and weaker flight ability in adult *B. impatiens* prey likely contributed to prolonged longevity and increased fecundity in adult *C. attenuata*. In this case, 12 to 24 adults of *B. impatiens* daily per predator were considered optimal prey density in the mass rearing of adult *C. attenuata* adult.

Author Contributions: Conceptualization, D.Z. and H.W.; methodology, D.Z., L.Z. and H.W.; software, W.X. and J.X.; validation, D.Z., T.A.C., L.Z., W.X., J.X. and H.W.; formal analysis, D.Z. and L.Z.; investigation, D.Z., M.W. and X.X.; resources, W.X. and H.W.; data curation, D.Z., M.W., X.X. and H.W.; writing—original draft preparation, D.Z.; writing—review and editing, T.A.C., L.Z., W.X., J.X. and H.W.; visualization, D.Z., T.A.C., L.Z., W.X., J.X. and H.W.; supervision, H.W.; project administration, D.Z., J.X. and X.X.; funding acquisition, D.Z., T.A.C., L.Z., W.X. and H.W. All authors have read and agreed to the published version of the manuscript.

Funding: This research was funded by Science and Technology Innovation Foundation for Young Scientists of Tianjin Academy of Agricultural Sciences (No. 201911), National Key Research and Development Program of China (No. 2017YFD0201000, 2017YFD0201707) and the USDA Agricultural Research Service project Insect Biotechnology Products for Pest Control and Emerging Needs in Agriculture (No. 5070-22000-037-00-D).

Institutional Review Board Statement: Not applicable.

Informed Consent Statement: Not applicable.

Data Availability Statement: Not applicable.

Acknowledgments: We thank Wan-Qi Xue (Shenyang Normal University, China) for helping in species identification.

Conflicts of Interest: The authors declare no conflict of interest. The funders had no role in the design of the study; in the collection, analyses or interpretation of data; in the writing of the manuscript or in the decision to publish the results. USDA is an equal opportunity provider and employer. Mention of trade names or commercial products in this publication is solely for the purpose of providing specific information and does not imply recommendation or endorsement by the USDA.

References

1. Cock, M.J.W. *Bemisia Tabaci, an Update 1986–1992 on the Cotton Whitefly with an Annotated Bibliography*; CAB International Institute of Biological Control: Ascot, UK, 1993; p. 78.
2. Gerling, D.; Alomar, O.; Arn, J. Biological control of *Bemisia tabaci* using predators and parasitoids. *Protection* **2001**, *20*, 779–799. [CrossRef]
3. Parrela, M.P. Biological control in protected culture: Will it continue to expand? *Phytoparasitica* **2008**, *3*, 3–6. [CrossRef]

4. Seabra, S.G.; Brás, P.G.; Martins, J.; Martins, R.; Wyatt, N.; Shirazi, J.; Rebelo, M.T.; Franco, J.C.; Mateus, C.; Figueiredo, E.; et al. Phylogeographical patterns in *Coenosia attenuata* (Diptera: Muscidae): A widespread predator of insect species associated with greenhouse crops. *Biol. J. Linnean Soc.* **2015**, *114*, 308–326. [CrossRef]

5. Hennig, W. Muscidae. In *Die Fliegen der Palaearktischen Region*; Lindner, E., Ed.; Schweizerbart'sche Verlagsbuchhandlung: Stuttgart, Germany, 1964; Volume 7.

6. Colombo, M.; Eördegh, F.R. Discovery of *Coenosia attenuata*, an active predator on aleyrodids, in protected crops in Liguria and Lombardia. *L'Informatore Agrar.* **1990**, *47*, 187–189.

7. Kühne, S. Open rearing of generalist predators: A strategy for improvement of biological pest control in greenhouses. *Phytoparasitica* **1998**, *26*, 277–281. [CrossRef]

8. Martinez, M.; Cocquempot, C. La mouche *Coenosia attenuata* nouvel auxiliaire prometteur en culture protégée. *PHM-Rev. Hortic.* **2000**, *414*, 50–52.

9. Rodríguez-Rodríguez, M.D.; Aguilera, A. *Coenosia attenuata*, una nueva mosca a considerar en el control biológico de las plagas hortícolas. *Phytoma España* **2002**, *141*, 27–34.

10. Hoebeke, E.R.; Sensenbach, E.J.; Sanderson, J.P. First report of *Coenosia attenuata* Stein (Diptera: Muscidae), an old world 'hunter fly' in North America. *Proc. Entomol. Soc. Wash.* **2003**, *105*, 769–775.

11. Pohl, D.; Uygur, F.N.; Sauerborn, J. Fluctuations in population of the first recorded predatory fly *Coenosia attenuata* in cotton fields in Turkey. *Phytoparasitica* **2003**, *31*, 446–449. [CrossRef]

12. Xue, W.Q.; Tong, Y.F. A taxonomic study on *Coenoia tigrina* species-group (Diptera: Muscidae) in China. *Entomol. Sin.* **2003**, *10*, 281–290.

13. Pohl, D.; Kühne, S.; Karaca, İ.; Moll, E. Review of *Coenosia attenuata* Stein and its first record as a predator of important greenhouse pests in Turkey. *Phytoparasitica* **2012**, *40*, 63–68. [CrossRef]

14. Ndiaye, O.; Ndiaye, S.; Djiba, S.; Ba, C.T.; Vaughan, L.; Rey, J.-Y.; Vayssières, J.-F. Preliminary surveys after release of the fruit fly parasitoid *Fopius arisanus* Sonan (Hymenoptera Braconidae) in mango production systems in Casamance (Senegal). *Fruit* **2015**, *70*, 91–99. [CrossRef]

15. Téllez, M.M.; Tapia, G. Acción depredadora de *Coenosia attenuata* Stein (Díptera: Muscidae) sobre otros enemigos naturales en condiciones de laboratorio. *Bol. San. Veg. Plagas* **2006**, *32*, 491–498.

16. Zou, D.Y. Current research status of key natural enemies in Tianjin. *Chin. Rur. Sci. Technol.* **2015**, *10*, 38–39.

17. Zou, D.Y.; Coudron, T.A.; Xu, W.H.; Gu, X.S.; Wu, H.H. Development of immature tiger-fly *Coenosia attenuata* (Stein) reared on larvae of the fungus gnat *Bradysia impatiens* (Johannsen) in coir substrate. *Phytoparasitica* **2017**, *45*, 75–84. [CrossRef]

18. Zou, D.Y.; Xu, W.H.; Liu, X.L.; Bai, Y.C.; Liu, B.M.; Xu, J.Y.; Hu, X.; Gu, X.S.; Wu, H.H. *Coenosia attenuata* Stein (Diptera: Muscidae): Research progress and prospects in biological control. *J. Environ. Entomol.* **2017**, *39*, 444–452.

19. Bautista-Martínez, N.; Illescas-Riquelme, C.P.; García-Ávila, C.J. First report of "hunter-fly" *Coenosia attenuata* (Diptera: Muscidae) in Mexico. *Fla. Entomol.* **2017**, *100*, 174–175. [CrossRef]

20. Solano-Rojas, Y.; Pont, A.; De Freitas, J.; Moros, G.; Goyo, Y. First record of *Coenosia attenuata* Stein, 1903 (Diptera: Muscidae) in Venezuela. *An. Biol.* **2017**, *39*, 223–226. [CrossRef]

21. Couri, M.S.; Sousa, V.R.; Lima, R.M.; Dias-Pini, N.S. The predator *Coenosia attenuata* Stein (Diptera, Muscidae) on cultivated plants from Brazil. *An. Acad. Bras. Ciênc.* **2018**, *90*, 179–183. [CrossRef]

22. Kühne, S.; Schrameyer, K.; Müller, R.; Menzel, F. Räuberische fliegen: Ein bisher wenig beachteter nützlingskomplex in Gewächshäusern. *Mitt. Biol. Bundesanst.* **1994**, *302*, 1–75.

23. Kühne, S.; Schiller, K.; Dahl, U. Beitrag zur lebensweise, morphologie und entwicklungsdauer der räuberischen fliege *Coenosia attenuata* Stein (Diptera: Muscidae). *Gesunde Pflanz.* **1997**, *49*, 100–106.

24. Moreschi, I.; Colombo, M. Una metódica per l'allevamento dei Ditteri predatori *Coenosia attenuata* e *C. strigipes*. *Inf. Fitopatol.* **1999**, *49*, 61–64.

25. Kühne, S. Räuberische Fliegen der Gattung *Coenosia* Meigen, 1826 (Diptera: Muscidae) und die Möglichkeit ihres Einsatzes bei der biologischen Schädlingsbekämpfung. *Stud. Dipterol.* **2000**, *9*, 78.

26. Prieto, R.; Figueiredo, E.; Mexia, A. *Coenosia attenuata* Stein (Diptera: Muscidae): Prospeccao e actividade em culturas protegidas em Portugal. *Bol. San. Veg. Plagas* **2005**, *31*, 39–45.

27. Sanderson, J.; Ugine, T.; Wraight, S.; Sensenbach, E. Something for nothing: The beneficial hunter fly may be helping you manage pests. *Growertalkers* **2009**, *73*, 74–76.

28. Sensenbach, E.J.; Wraight, S.P.; Sanderson, J.P. Biology and predatory feeding behavior of larvae of the hunter fly *Coenosia attenuata*. *IOBC/WPRS B.* **2005**, *28*, 229–232.

29. Pinho, V.; Mateus, C.; Rebelo, M.T.; Kühne, S. Distribuição espacial de *Coenosia attenuata* Stein (Diptera: Muscidae) e das suas presas em estufas de hortícolas na região Oeste, Portugal. *Bol. San. Veg. Plagas* **2009**, *35*, 231–238.

30. Téllez, M.D.M.; Tapia, G.; Gámez, M.; Cabello, T.; van Emden, H.F. Predation of *Bradysia* sp. (Diptera: Sciaridae), *Liriomyza trifolii* (Diptera: Agromyzidae) and *Bemisia tabaci* (Hemiptera: Aleyrodidae) by *Coenosia attenuata* (Diptera: Muscidae) in greenhouse crops. *Eur. J. Entomol.* **2009**, *106*, 199–204. [CrossRef]

31. Kühne, S.; Heller, K. Sciarid fly larvae in growing media—Biology, occurrence, substrate and environmental effects and biological control measures. In *Peat in Horticulture—Life in Growing Media*; Schmilewski, G., Ed.; International Peat Society: Amsterdam, The Netherlands, 2010; pp. 95–102.

32. Ugine, T.A.; Sensenbach, E.J.; Sanderson, J.P.; Wraight, S.P. Biology and feeding requirements of larval hunter flies *Coenosia attenuata* (Diptera: Muscidae) reared on larvae of the fungus gnat *Bradysia impatiens* (Diptera: Sciaridae). *J. Econ. Entomol.* **2010**, *103*, 1149–1158. [CrossRef]

33. Martins, J.; Domingos, C.; Nunes, R.; Garcia, A.; Ramos, C.; Mateus, C.; Figueiredo, E. *Coenosia attenuata* (Diptera: Muscidae), um predador em estudo para utilização em culturas protegidas. *Rev. Ciênc. Agrár.* **2012**, *35*, 229–235.

34. Mateus, C. Bioecology and behaviour of *Coenosia attenuata* in greenhouse vegetable crops in the Oeste region, Portugal. *Bull. Insectol.* **2012**, *65*, 257–263.

35. Zou, D.Y.; Zhang, L.S.; Wu, H.H.; Gu, X.S.; Xu, W.H.; Liu, X.L. Population dynamics of *Coenosia attenuata* and prey in Tianjin. In *Green Plant Protection and Rural Revitalization*; Chen, W.Q., Zheng, C.L., Wen, L.P., Ni, H.X., Feng, L.Y., Eds.; China Agricultural Science and Technology Press: Beijing, China, 2018; pp. 171–172.

36. Fabian, S.T.; Sumner, M.E.; Wardill, T.J.; Rossoni, S.; Gonzalez-Bellido, P.T. Interception by two predatory fly species is explained by a proportional navigation feedback controller. *J. R. Soc. Interface* **2018**, *15*, 20180466. [CrossRef]

37. Bonsignore, C.P. Environmental factors affecting the behavior of *Coenosia attenuata*, a predator of *Trialeurodes vaporariorum* in tomato greenhouses. *Entomol. Exp. Appl.* **2016**, *158*, 87–96. [CrossRef]

38. Martins, J.; Mateus, C.; Ramos, A.; Figueiredo, E. An optimized method for mass rearing the tiger-fly, *Coenosia attenuata* (Diptera: Muscidae). *Eur. J. Entomol.* **2015**, *112*, 470–476. [CrossRef]

39. Zou, D.Y.; Coudron, T.A.; Xu, W.H.; Xu, J.Y.; Wu, H.H. Performance of the tiger-fly *Coenosia attenuata* Stein reared on the alternative prey, *Chironomus plumosus* (L.) larvae in coir substrate. *Phytoparasitica* **2021**, *49*, 83–92. [CrossRef]

40. Morris, D.E.; Cloutier, C. Biology of the predatory fly *Coenosia tigrina* (Fab.) (Diptera: Anthomyiidae): Reproduction, development, and larval feeding on earthworms in the laboratory. *Can. Entomol.* **1987**, *119*, 381–393. [CrossRef]

41. Moreschi, I.; Süss, L. Osservazioni biologiche ed etologiche su *Coenosia attenuata* Stein e *Coenosia strigipes* Stein (Diptera Muscidae). *Boll. Zool. Agrar. Bachicol.* **1998**, *30*, 185–197.

42. Gilioli, G.; Baumgärtner, J.; Vacante, V. Temperature influences on functional response of *Coenosia attenuata* (Diptera: Muscidae) individuals. *J. Econ. Entoml.* **2005**, *98*, 1524–1530. [CrossRef]

43. Gonzalez-Bellido, P.T.; Wardill, T.J.; Juusola, M. Compound eyes and retinal information processing in miniature dipteran species match their specific ecological demands. *PNAS* **2011**, *108*, 4224–4229. [CrossRef] [PubMed]

44. Song, Z.; Juusola, M. Refractory sampling links efficiency and cost of sensory encoding to stimulus statistics. *J. Neurosci.* **2014**, *34*, 7216–7237. [CrossRef]

45. Wardill, T.J.; Gonzalez-Bellido, P.T.; Tapia, G.; Peng, H.; Olberg, R.M. The miniature dipteran killer fly *Coenosia attenuata* exhibits adaptable aerial prey capture strategies. In Proceedings of the International Conference on Invertebrate Vision, Fjälkinge, Sweden, 1–8 August 2013. [CrossRef]

46. Sabeli, M.W. Predatory arthropods. In *Natural Enemies*; Crawley, M.J., Ed.; Blackwell Scientific: Oxford, UK, 1992; pp. 225–264.

47. Giambiasi, M.; Rodríguez, A.; Arruabarrena, A.; Buenahora, J. First report of *Coenosia attenuata* (Stein, 1903) (Diptera, Muscidae) in Uruguay, confirmed by DNA barcode sequences. *Check List* **2020**, *16*, 749–752. [CrossRef]

48. Chen, M.Y.; Liu, H.P.; Cheng, J.; Chiang, S.Y.; Liao, W.P.; Lin, W.Y. Transgenerational impact of DEHP on body weight of *Drosophila*. *Chemosphere* **2019**, *221*, 493–499. [CrossRef] [PubMed]

49. Wilkinson, J.D.; Daugherty, D.M. The biology and immature stages of *Bradysia impatiens* (Diptera: Sciaridae). *Ann. Entomol. Soc. Am.* **1970**, *63*, 656–660. [CrossRef]

Article

Horizontal Honey-Bee Larvae Rearing Plates Can Increase the Deformation Rate of Newly Emerged Adult Honey Bees

Juyeong Kim [1], Kyongmi Chon [1,*], Bo-Seon Kim [1], Jin-A Oh [1], Chang-Young Yoon [1], Hong-Hyun Park [1] and Yong-Soo Choi [2]

[1] Toxicity and Risk Assessment Division, Department of Agro-food Safety and Crop Protection, National Institute of Agricultural Sciences, Rural Development Administration, Wanju-gun 55365, Korea; kjy.sara@gmail.com (J.K.); kbs9249@naver.com (B.-S.K.); oja5074@korea.kr (J.-A.O.); evermoo2600@korea.kr (C.-Y.Y.); honghyunpark@korea.kr (H.-H.P.)

[2] Sericulture and Apiculture Division, Department of Agricultural Biology, National Institute of Agricultural Sciences, Rural Development Administration, Wanju-gun 55365, Korea; beechoi@korea.kr

* Correspondence: kmchon6939@korea.kr

Simple Summary: Rearing honey bee (*Apis mellifera*) larvae in vitro is an important method for studying bee larvae diseases or the toxicity of pesticides on bees. Laboratory experiments for bee larvae are usually performed by placing a rearing plate horizontally during all developmental stages. However, recent studies have demonstrated that a horizontal rearing environment can cause the deformation of emerged bees. Most studies adopted a vertical rearing method to reduce such deformation, but there is a lack of information on the emergence rates and deformation rates of bees reared on vertical or horizontal plates. Therefore, in this study, we examined the effect of placing the plates vertically and horizontally on newly emerged bees. There were no significant differences in larval mortality, pupal mortality, and adult emergence rates between horizontal and vertical rearing plates. However, the adult deformation rates of the horizontal plates were significantly higher than those of the vertical plates. In conclusion, we suggest that the vertical rearing method is more suitable when considering the deformation rate of the control group to verify the sublethal effects of pesticides on honey bees.

Abstract: Rearing honey bee larvae in vitro is an ideal method to study honey bee larval diseases or the toxicity of pesticides on honey bee larvae under standardized conditions. However, recent studies reported that a horizontal position may cause the deformation of emerged bees. Accordingly, the purpose of this study was to evaluate the emergence and deformation rates of honey bee (*Apis mellifera ligustica*) larvae reared in horizontal and vertical positions. The study was conducted under the same laboratory conditions with three experimental groups, non-capped or capped horizontal plates and capped vertical plates. However, our results demonstrated that the exhibited adult deformation rates of the horizontal plates were significantly higher (27.8% and 26.1%) than those of the vertical plates (11.9%). In particular, the most common symptoms were deformed wings and an abnormal abdomen in the horizontal plates. Additionally, adults reared on horizontal plates were substantially smaller (10.88 and 10.82 mm) than those on vertical plates (11.55 mm). Considering these conclusions, we suggest that a vertical rearing method is more suitable when considering the deformation rates of the control groups to verify the sublethal effects of pesticides on honey bees.

Keywords: *Apis mellifera*; deformation; emergence; honey bee; in vitro rearing; larvae

Citation: Kim, J.; Chon, K.; Kim, B.-S.; Oh, J.-A; Yoon, C.-Y.; Park, H.-H.; Choi, Y.-S. Horizontal Honey-Bee Larvae Rearing Plates Can Increase the Deformation Rate of Newly Emerged Adult Honey Bees. *Insects* **2021**, *12*, 603. https://doi.org/10.3390/insects12070603

Academic Editors: Man P. Huynh, Kent S. Shelby and Thomas A. Coudron

Received: 6 June 2021
Accepted: 23 June 2021
Published: 1 July 2021

1. Introduction

Many studies have reported recent significant pollinator declines and increased honey bee (*Apis mellifera*) colony losses in many countries [1–5]. Several stressing factors, such as pathogens, climate change, parasites, habitat loss, lack of nutrition, pesticides, and diseases are considered to explain the decline and colony losses [6–9]. However, a single causative

Insects **2021**, *12*, 603. https://doi.org/10.3390/insects12070603   https://www.mdpi.com/journal/insects

stressor factor has not been conclusively identified, because of the complexity related to concurrent multiple stressors [4,10–12]. Among the several factors suggested, pesticides are regarded as one of the most crucial causes of adverse honey-bee health and colony declines [4,13–15]. Various studies have been conducted to determine the exposure effects of pesticides on adult honey bees, therefore, standard methods for investigating the effects of pesticides on adult bees have been well investigated both in vivo and in vitro [15,16]. Nevertheless, in contrast to controllable laboratory conditions, field experiments in hives are impacted by numerous uncontrollable factors such as season, colony genetic variation, climate, and resource availability [17–19]. Because of these uncontrolled variables, the in vitro procedure of rearing honey-bee larvae has been proposed to evaluate the toxicity of pesticides on honey-bee broods (larvae, pupae, and adults) [11]. Rearing larvae in vitro is a practical protocol to study larval pathogens, development, and caste differentiation in honey bees [20–23]. Nevertheless, available data in the publications concerning the lethal and sublethal effects on honey-bee larvae are rather poor compared to adult bees [11,24,25].

In 1933, the first informative report investigating the caste differentiation of queen and worker bees, as well as hand-feeding bee larvae with a diet containing royal jelly in vitro, was published [26]. The larval diet composition was improved and optimized by Rembold and Lackner [23], and Vandenberg and Shimanuki [27]. Vandenberg and Shimanuki [27] further developed methods of rearing one larva per cup and feeding them the correct quantity of diet daily. Wittmann and Engels [28] reported an in vitro rearing method as a risk assessment tool to study the toxicity of pesticides. Davis et al. [29] provided diets containing carbofuran and dimethoate to larvae reared in the laboratory, and Peng et al. [21] utilized rearing honey-bee larvae in vitro for assessing the toxicity of pesticides on honey bees. Most notably, seven laboratories across five different countries performed ring tests according to the improved in vitro methods to assess LD_{50} for acute toxicity of dimethoate in 2005 and 2008 [17]. The ring test participants achieved adult emergence rates greater than 80% in 43% of their control trials and greater than 90% in 17% of their control trials [12,17]. The Organization for Economic Co-operation and Development (OECD) guidelines for the honey-bee larvae toxicity tests under laboratory conditions were published in 2013 (single exposure) and 2016 (repeated exposure), based on the methods that were developed in ring tests [30,31].

Larval mortality and pupal mortality in the natural hive occurred at approximately 15% [32,33]. Therefore, the OECD guidelines specify that the total mortality during larval and pupal developmental stages should not exceed 15% in the controls, otherwise the study is considered invalid [30,31]. Namely, the mortalities of the control groups should be considered for validation of the test [32].

Accordingly, researchers have focused on increasing the survival rate of the larval stage, and there are several examples of these methods. They mainly improved the survival rate of larvae according to the composition of larval food, quality of royal jelly, larval age at grafting, rearing conditions (temperature and relative humidity) in the laboratory, or reducing contact between larvae and fecal materials using absorbents such as Kimwipes (filter paper) [12,32,34–36].

However, even if in vitro rearing protocols have been improved over the years, variable (inconsistent) survival rates in the controls of each laboratory have been reported continuously [12,17]. Zhu et al. [37] reported the larval mortalities of controls were approximately 17.5% at D6 after grafting. Additionally, low emergence rates of controls (≤50%) were noted in several experimental studies based upon Aupinel et al. [11,12]. Namely, this means that several research institutes have already performed a larval toxicity test, with control mortalities higher than the OECD guidelines [17,18,22,38]. The inconsistent results across different laboratories may reflect subtle differences in the brood sources and the laboratory conditions, or be related to the effects of the mechanical stress of grafting [12,19,39]. It is also a more common practice to set up the rearing using 48-well tissue plates, with the grafting cell cups placed horizontally, which is according to the OECD guideline (2016). However, some studies have mentioned that the horizontal rearing positions during the developmental stages may lead to the deformation of wings and abdomen (humpback) in emerged bees, because the larvae or pupae can then withstand abnormal vertical positions [32,40,41]. Riessberger-Gallé et al. [40] proposed a method to prevent these deformations, in which a 48-well tissue plate was sealed with a thin wax layer and set vertically as in natural beehives. However, presently, specific information on the emergence rates and deformation rates of the newly emerged honey bees when the rearing plates are placed vertically or horizontally has not yet been reported. Therefore, in this study, we aimed to evaluate the specific difference in the emergence rate and deformation rate of emerged bees when the rearing plates were positioned horizontally or vertically during the pupal developmental stage.

2. Materials and Methods

2.1. Rearing Honey Bee Larvae In Vitro

The honey-bee larvae (*Apis mellifera ligustica*) were randomly obtained from three healthy colonies at an apiary located at the National Institute of Agricultural Sciences in Korea (35°49′47.0″ N, 127°02′26.0″ E). The source colonies and bees were not treated against *Varroa destructor* for four weeks prior to the experiment. The honey-bee queens were caged on wax combs using queen excluders to lay eggs. The freshly laid eggs were confirmed the next day (24 h after the queens were caged), and after that, the combs containing the hatched first instar larvae (72 h after the queens laid the eggs) were delivered to the laboratory for grafting. Before grafting, the grafting cell cups (Nicotplast, Maisod, France) were disinfected with 70% ethanol and were then used, after UV sterilization for 30 min in a laminar-flow hood. The larval rearing procedure followed OECD No. 239 [31]. On day 1 (D1), 20 μL of larval diet A was loaded into the grafting cell cup, and healthy first instar larvae were transferred into the cell cups of 48-well tissue plates (SPL, Pocheon-si, Korea). During grafting, a clay pack (Caremate, Hwaseong-si, Korea) was preheated in the microwave and the pack was laid underneath 48-well tissue plates to minimize the temperature effect on the larvae [32]. After grafting, the 48-well tissue plates were placed horizontally in a sealed desiccator (Nalgene, Rochester, NY, USA) in a constant temperature incubator (DAIHAN Scientific Co., Wonju-si, Korea) maintained at 35 °C and 95 ± 5% relative humidity (RH) using a saturated solution of potassium sulfate (Junsei, Tokyo, Japan) during the larval stages (D1–D8). The pupal stages (D8–D15) were maintained at 80 ± 5% RH using a saturated solution of sodium chloride (Sigma–Aldrich, St. Louis, MO, USA). In the emergence stages (D15–D21), the 48-well tissue plates were transferred individually into emergence boxes with a 50% sucrose solution, and placed in the incubator to maintain 50% RH and 35 °C.

For larval feeding, d-glucose (Difco, Sparks, NV, USA) and d-fructose (Junsei, Japan) were added to water filtered with a 0.20 μm filter (Sartorius, Göttingen, Germany), and then yeast extract (Bacto, Sparks, NV, USA) was added and mixed. Lastly, the solution was mixed with royal jelly (Haechangol Honey Farm, Yeongwol, Korea) [12]. Following the OECD guidelines, the larval diets included a total of 160 μL of each standardized volume during the six days (excluding D2) of the larval stages, where the larval diet volume and components have been summarized in Table 1. Before feeding the larvae, diets were preheated in an incubator maintained at 35 °C.

Table 1. The volume and larval diet component percentages, according to the OECD guidelines [30,31].

Day	1	3	4, 5, 6
Volume of diet/larva (µL)	20	20	30, 40, 50
Diet component	Diet A	Diet B	Diet C
Royal jelly (%)	50.0	50.0	50.0
Distilled water (%)	37.0	33.5	30.0
Glucose (%)	6.0	7.5	9.0
Fructose (%)	6.0	7.5	9.0
Yeast extract (%)	1.0	1.5	2.0

2.2. Experimental Design

The non-capped and horizontally oriented groups (NHG) were placed in 48-well tissue plates without a wax layer and set horizontally, and the capped and horizontally oriented groups (CHG) were placed in 48-well tissue plates that were capped with the artificial wax layers and set horizontally. Meanwhile, the capped and vertically oriented groups (CVG) were placed in 48-well tissue plates that were capped with the artificial wax layers and set vertically. All groups were not treated with any chemical reagents. The experiments of each group were tested with 4 replicates (36 larvae per plate). Artificial wax layers were prepared by dissolving 4.0 g of pure beeswax. The size of the wax layers was 14 cm × 10 cm × 0.4 mm. In each grafting cell cup of the plate, small orifices were made to allow air exchange. Rearing plates were sealed with the perforated wax layers on D15 after grafting, particularly vertical plates (CVG) placed carefully upright so pupae were facing towards the opening.

2.3. Mortality and Abnormal Symptoms

Mortality and abnormal symptoms at each developmental stage were visually observed and recorded every day. After setting the survival rates of the larvae at 100% on D3, larval mortality, and abnormal symptoms were monitored as early death and melanizing death from D4 to D8 (larval stages) [12,30,31]. The larvae were considered as dead when the larval color became dark or they had no motion, and were removed daily from the test plates. On D7, no additional diet was fed, and on D8, the number of larvae with uneaten diets was recorded. From D8 to D21 (pupal stages), pupal mortality was assessed based on failed molt and failed adult molt [12,31,42]. From D16 to D21 (emergence stages), the number of newly emerged bees was observed and recorded daily. Individuals that died after emergence or fully developed bees that stayed in the cell without breaking the wax layer were considered to be emerged bees. All deformation symptoms and morphological characteristics (weight and length) of dead adult bees after emerging were observed and measured immediately. After all deformation symptoms of living emerged bees were observed on D21, the weight and length of the emerged bees were measured. The whole-body length of an adult bee was measured from the tip of its head to the tip of its abdomen. In particular, when a humpback was present in the bee, the total length of the body was measured as it was in the unstretched state. The symptoms of newly emerged adult bees were classified as surviving normal (SN), deformed wings (DW), deformed antennae (DA), and abnormal abdomen shape (AAS) (Table 2) [43]. The larval mortality, pupal mortality, adult emergence rate, and deformation rate were calculated for each group using the following formulae [31]:

Larval mortality = (the number of dead larvae from D3 to D8/the number of larvae on D3) × 100

Pupal mortality = (the number of dead pupae from D8 to D21/the number of pupae on D8) × 100

Adult emergence rate = (the number of emerged bees/the number of larvae on D3) × 100

Deformation rate = (the number of deformed bees/the number of emerged bees) × 100

Table 2. List and descriptions of mortality symptoms during developmental stage and deformation symptoms observed in emerged adult bees. This table is modified from Fine et al. [42] and Barbosa et al. [43].

Symptom	Description
Early death (ED)	Sunk in diet, failed to maintain the C-shape, flattening
Melanizing death (MD)	Death with darkening internally or externally, having black spots
Failed molt (FM)	Failure to evert imaginal discs, but the pupal molt is incomplete
Failed adult molt (FA)	Failure to emerge from final molt
Surviving normal (SN)	Survived and successfully eclosed as bees
Deformed wings (DW)	Eclosed as bees with deformed wings
-with short wings (DSW)	Eclosed as bees with short wings
-with tangled wings (DTW)	Eclosed as bees with tangled wings
Deformed antennae (DA)	Eclosed as bees with deformed antennae
Abnormal abdomen shape (AAS)	Eclosed as bees with abnormal abdomen shape (humpback)

2.4. Statistical Analysis

A statistical analysis of the data was carried out using the SPSS statistical software program (SPSS 20.0 Inc., Chicago, IL, USA). A Pearson's chi-square test was used to compare larval mortality, pupal mortality, and adult emergence rates among the three groups. A Fisher's exact test was used to assess differences between the total deformation rates of emerged bees in the three groups. The Kaplan–Meier log-rank test was used to compare the survival curves of each group. The mortality, emergence rates, and deformation rates were expressed as means ± SE. Means ± SE (standard error) were calculated for the four replicate values of each group. The emergence date, adult weight, and length were expressed as means ± SD (standard deviation). The dates of emergence, and the adult weight and length at D21, were tested by the one-way ANOVA and were determined using Tukey's HSD test to compare the values among the three groups. A p-value of <0.05 was considered as a statistically significant difference.

3. Results

3.1. In Vitro Mortality, Adult Emergence Rates, and Survival

Larval mortality and pupal mortality means ± SE were as follows: 4.9 ± 0.7% and 16.1 ± 2.8%, respectively, in the NHG; 7.7 ± 4.1% and 13.5 ± 3.4%, respectively, in the CHG; and 4.2 ± 2.7% and 15.3 ± 2.2%, respectively, in the CVG. The three groups satisfied the OECD test condition that the larval mortalities were less than 15% in the negative controls. On D21, total emergence rates were 79.9 ± 3.3% in the NHG, 79.9 ± 6.7% in the CHG, and 81.3 ± 0.9% in the CVG, which corresponds to the OECD test condition that the adult emergence rate should be ≥70% in the controls. No statistically significant differences were detected among the three groups with respect to larval mortality, pupal mortality, and adult emergence rates (chi-square test, $p > 0.05$, Table 3). The survival curves of the honey-bee larvae have been illustrated in Figure 1, where no significant differences were found among the three groups in terms of their survival (Kaplan–Meier log-rank test, $p > 0.05$).

Table 3. Larval mortality, pupal mortality, and adult emergence rate in *Apis mellifera*. Values are means ± SE. No significant differences in mortalities and emergence rates were found among the three groups (chi-square test, $p > 0.05$).

	NHG [1]	CHG [2]	CVG [3]	*p* Values
Larval mortality (%)	4.9 (±0.7)	7.7 (±4.1)	4.2 (±2.7)	0.396
Pupal mortality (%)	16.1 (±2.8)	13.5 (±3.4)	15.3 (±2.2)	0.843
Adult emergence rate (%)	79.9 (±3.3)	79.9 (±6.7)	81.3 (±0.9)	0.943

[1] NHG: the non-capped and horizontally oriented groups; [2] CHG: the capped and horizontally oriented groups; [3] CVG: the capped and vertically oriented groups.

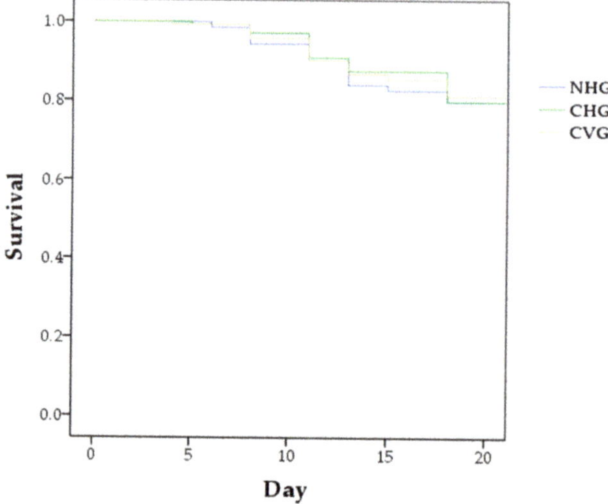

Figure 1. The survival curves of the honey-bee larvae of the horizontally oriented groups and the vertically oriented groups over 21 days, which were assessed by the Kaplan–Meier log-rank test ($p > 0.05$). NHG: the non-capped and horizontally oriented groups; CHG: the capped and horizontally oriented groups; CVG: the capped and vertically oriented groups.

3.2. Emergence Rates by Time (Days)

The emergence rates of each group were assessed according to the time (days). By D17, honey bees from the groups had not emerged, and then, on D18, worker bees began to emerge at 4.0% in the NHG, 0.7% in the CHG, and 1.4% in the CVG. On D19, more than half of the bees (75.7%) emerged in the NHG, and 59.0% in the CVG, compared to the 39.6% of bees that emerged in the CHG. Finally, on D21, 79.9% of bees emerged in the NHG, 79.9% of bees emerged in the CHG, and 81.3% of bees emerged in the CVG (Figure 2). The mean emergence date of the NHG was 19.00 ± 0.32 days, that of the CHG was 19.85 ± 0.93 days, and that of the CVG was 19.39 ± 0.74 days. The three groups demonstrated a statistically significant difference in emergence date (Tukey's HSD test, $p < 0.05$).

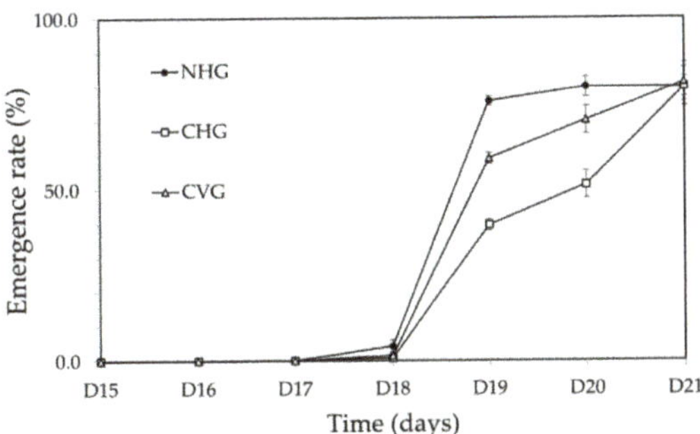

Figure 2. Emergence rates of each group from D15 to D21. NHG: the non-capped and horizontally oriented groups; CHG: the capped and horizontally oriented groups; CVG: the capped and vertically oriented groups.

3.3. Deformation Rate of Newly Emerged Adult Bees

The total deformation rates were $27.8 \pm 7.7\%$ in the NHG, $26.1 \pm 6.9\%$ in the CHG, and $11.9 \pm 2.0\%$ in the CVG. The deformation rates of NHG and CHG were significantly higher than those of CVG (Fisher's exact test, $p < 0.05$, Figure 3). Figure 4 illustrates examples of normal and deformed bees in the three groups. DW was shorter or had tangled ends compared to the normal wings. DA had curved ends compared to the normal antennae. Additionally, an AAS was more curved compared to the normal abdomen shape. Deformation symptoms were classified into 9 categories, including AAS, DSW, DTW, deformed with tangled wings asymmetrically (DTWA), and DA. In the NHG, the deformation rates were AAS (13.9%), DSW (0.9%), DTW (0.9%), DTWA (3.5%), AAS + DW (4.3%), DW + DA (0.9%), and AAS + DW + DA (2.6%). In the CHG, the deformation rates were AAS (4.3%), DSW (4.3%), DTW (3.5%), DTWA (6.1%), DA (0.9%), AAS + DW (4.3%), AAS + DA (1.7%), DW + DA (0.9%), and AAS + DW + DA (0.9%). In CVG, the deformation rates were AAS (3.4%), DSW (1.7%), DTW (1.7%), DTWA (3.4%), DA (0.9%), and DW + DA (0.9%) (Table 4).

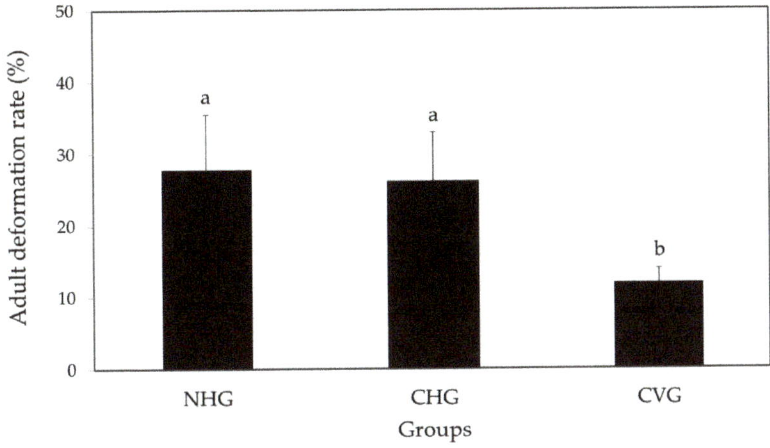

Figure 3. Deformation rates of newly emerged adult bees. The different letters above the bars indicate significant differences among the three groups (Fisher's exact test, $p < 0.05$). NHG: the non-capped and horizontally oriented groups; CHG: the capped and horizontally oriented groups; CVG: the capped and vertically oriented groups.

Figure 4. The observed symptoms of newly emerged adult bees. (**A**) Surviving normal (SN); (**B**) Deformed with short wings (DSW); (**C**) Deformed with tangled wings (DTW); (**D**) Normal (above) and abnormal abdomen shape (AAS) (below); (**E**) Deformed antennae (DA). Scale bar = 1 mm.

Table 4. The overall percentage of observed symptoms of newly emerged adult bees. Each percentage in the following table is a calculated value of the ratio of the number of deformed bees to the total number of emerged bees.

Observed Symptoms [1]	NHG [2]	CHG [3]	CVG [4]
SN (%)	73.0	73.0	88.0
AAS (%)	13.9	4.3	3.4
DSW (%)	0.9	4.3	1.7
DTW (%)	0.9	3.5	1.7
DTWA (%)	3.5	6.1	3.4
DA (%)	0.0	0.9	0.9
AAS + DW (%)	4.3	4.3	0.0
AAS + DA (%)	0.0	1.7	0.0
DW + DA (%)	0.9	0.9	0.9
AAS + DW + DA (%)	2.6	0.9	0.0

[1] SN: survived and successfully eclosed as bees; AAS: abnormal abdomen shape; DSW: deformed with short wings; DTW: deformed with tangled wings; DTWA: deformed with tangled wings asymmetrically; DA: deformed antennae; DW: deformed wings; [2] NHG: the non-capped and horizontally oriented groups; [3] CHG: the capped and horizontally oriented groups; [4] CVG: the capped and vertically oriented groups.

The main deformation symptoms were simplified into three categories as follows: AAS, DW and DA. The percentage of SN, AAS, DW, and DA were 73.1%, 17.0%, 8.7%, and 1.3%, respectively, in the NHG; 73.0%, 7.7%, 16.8%, and 2.5%, respectively, in the CHG; and 88.0%, 3.4%, 7.3% and 1.3%, respectively, in the CVG (Figure 5).

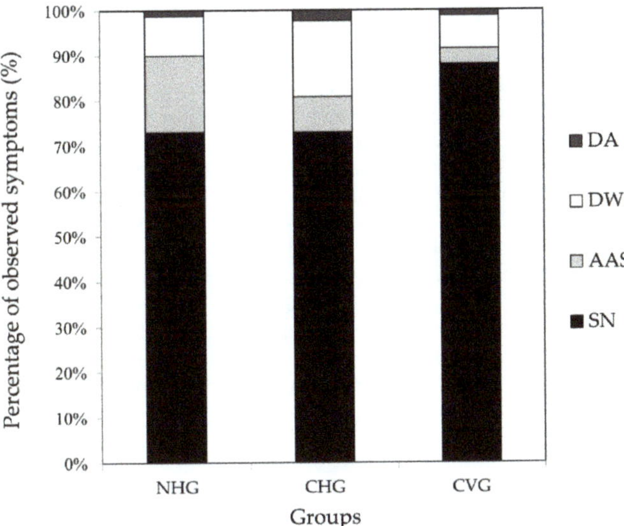

Figure 5. Percentage of observed symptoms of emerged bees. SN: survived and successfully eclosed as bees; AAS: abnormal abdomen shape; DW: deformed wings; DA: deformed antennae.

3.4. Body Weight and Length of Newly Emerged Adult Bees

The mean weight and length of the emerged bees were 67.40 ± 12.24 mg and 10.88 ± 0.96 mm in the NHG, 69.77 ± 12.63 mg and 10.82 ± 0.92 mm in the CHG, and 71.95 ± 12.12 mg and 11.55 ± 1.00 mm in the CVG, respectively. Adult weights of CVG were significantly higher than those of NHG ($p = 0.014$). Additionally, the length of the emerged bees in the CVG was significantly larger than that of the NHG and CHG (one-way ANOVA, $F_{(2, 344)} = 21.072$, $p < 0.05$, Table 5).

Table 5. The body weight and length of the newly emerged adult bees. Values are means ± SD. Means followed by the different letters across a row are significantly different (one-way ANOVA, $p < 0.05$).

	Weight (mg)	Length (mm)
NHG [1]	67.40 (±12.24) a	10.88 (±0.96) a
CHG [2]	69.77 (±12.63) ab	10.82 (±0.92) a
CVG [3]	71.95 (±12.12) b	11.55 (±1.00) b

[1] NHG: the non-capped and horizontally oriented groups; [2] CHG: the capped and horizontally oriented groups; [3] CVG: the capped and vertically oriented groups.

4. Discussion

Many studies have emphasized the importance of rearing honey-bee larvae in vitro for testing the toxicity of pesticides and, sequentially, experimental rearing methods have been systematically developed [12,18,32,44]. Natural or commercial hives consist of several honeybee combs with vertical structures for brood rearing and storing honey-bee products [45]. Conversely, as displayed in the OECD guidelines [30,31], many laboratories have generally performed experiments placing the rearing plates horizontally during all developmental stages (from larvae to adults). However, the horizontal rearing plates may induce deformations in emerged bees [32,41]. The purpose of this study was to analyze the effects of vertical and horizontal rearing plates on the emergence rates and deformation rates of newly emerged bees.

Honey bee larval defecation is usually on D7 after grafting [21]. Several recent studies suggest that the larvae should be transferred to a new clean plate, since larval mortality may increase due to the defecation of the larvae [12,17,38]. However, transferring larvae may cause mechanical stress or contamination of the larvae [39]. Besides, when feeding the total diet of 160 µL, the larvae eat all the food provided, thus, it is not necessary to move the larvae to new cell cups or clean the grafting cell cups [32]. Brodschneider et al. [41] placed rearing plates capped with thin wax layers vertically on D11. As a result, the total mortality until emergence was 16.3% in the control groups; they also reported similar flight performance between reared bees in vitro and hives. Likewise, Krainer et al. [46] demonstrated a total mortality of 28.1% in the control groups. They sealed rearing plates with the wax layer and the plates were placed vertically on D12. In our experiments, there were no significant differences in larval mortality, pupal mortality, and adult emergence rates among the three groups. The total adult emergence rates were about 80% in the three groups. Thus, each group exhibited similar survival rates of bee larvae regardless of plate position (horizontal and vertical).

For the most part, the worker bees reared in vitro emerged on D17–D18 after grafting [20]. In our experiments, bees of each group began to emerge from D18. Bees in the NHG were the earliest to emerge among the three groups. These differences could be due to external stimulus by the newly emerged bees that roam and stimulate other non-emergent pupae in the cells of 48-well tissue plates. Namely, bees in CHG and CVG may emerge more slowly, since the wax layer of the plate could interrupt this external stimulus. In particular, the CHG was the slowest among the three groups. This may be because more force is needed to break through the wax layer against gravity on the horizontal plate than on the vertical plate.

Tehel et al. [47] inoculated the honey-bee pupa with deformed wing virus (DWV) and tested the relative effects of the genotype of DWV on the mortality and wing malformation of adult honey bees. They placed the plate vertically so that pupae were horizontal in the incubators and monitored the pupal development. They observed that 23% of emerged bees in the control had wing deformities. In our experiments, the deformations of adults included deformed wings, deformed antennae, and abnormal abdomen shape, and horizontally oriented groups (NHG and CHG) demonstrated higher deformation rates (27.8% and 26.1%) than the vertically oriented groups (11.9%). In particular, the NHG and CHG had more wing or abdominal deformations than the CVG. Although the cause of deformation in emerged bees is not clear, all adults with deformations were derived only from deformed pupae [43]. Additionally, the wings of honey bees are formed during the pupal development stage [48]. When the rearing plates were horizontal, the pupae in the cell cups hold a vertical state. Thus, the pupal body is pulled down by gravity, affecting the wing and abdominal development [41]. In this regard, deformed bees may already have external deformations from the pupal stage.

In other studies, the rearing plates were capped and set vertically on D11 after grafting when pupation started, but in our experiments, the plates were capped with a wax layer on D15 to observe the mortality during the pupal stages. For this reason, it is thought that the horizontal condition between D11 and D14 (before capping the plates with wax layer) had already affected the pupae, resulting in abdominal and wing deformations in the CVG. Mechanisms for explaining antennae deformations due to physical external deformation have not been described in other studies.

The average adult weight ranged from 67.40 to 71.95 mg in our experiments; similarly, Brodschneider et al. [41] measured 76.6 ± 11.6 mg in emerged bees of the control groups. The average adult weight of the CVG was significantly higher than that of the NGH. The emerged bees of the CVG were significantly larger by 0.7 mm than the NHG and CHG, and this difference appeared to be due to the abnormal shapes, such as the humpback and abdominal shrinkage that were observed in bees reared horizontally.

Barbosa et al. [43] reported that when azadirachtin and spinosad were treated on the stingless bee, *Melipona quadrifasciata*, deformed pupae and emerged bees with wing, antennae, and leg deformities occurred. Additionally, they reported that deformed bees had side effects regarding flight activity or olfactory activity. Therefore, the deformations occurring in emerged bees have the potential to directly affect the activity of worker bees, and thus, these deformations can also be evaluated as sublethal effects [43]. In the present study, the mean adult deformation rates in CVG were approximately 12%. Thus, the vertical rearing method can be supported as a more appropriate method to verify the effects of pesticides on honey bees by considering the deformation rates in the control group. In future studies, the mechanisms of deformations in emerged honeybees that were identified here should be investigated.

5. Conclusions

Overall, there were no statistically significant differences in the emergence rates of adult bees between the horizontal and vertical plates, but the total deformation rates of the horizontal plates were significantly higher than those of the vertical plates. Our results are the first to discuss the emergence rates and deformation rates of honey bees concerning the position of plates in a laboratory. Considering these conclusions, the vertical rearing method with lower adult deformation rates appears to be more suitable, when considering the deformation rates of the control groups in order to verify the sublethal effects of pesticides on the bees. In the honey-bee larval toxicity test, according to the OECD guidelines, it is necessary to confirm the pupal mortality at the pupal stages (from D8 to D21). However, the vertical rearing plates must be capped with a wax layer on D15, so it may be difficult to check the pupal mortality after D15. Consequently, the rearing conditions and the position of rearing plates should be carefully considered depending on the purpose of the larval toxicity tests.

Author Contributions: Conceptualization, J.K. and K.C.; methodology, J.K. and K.C.; validation, J.K. and K.C.; formal analysis, J.K.; investigation, J.K., B.-S.K. and C.-Y.Y.; funding acquisition, K.C.; resources, Y.-S.C.; data curation, J.K.; writing—original draft preparation, J.K.; writing—review and editing, K.C.; visualization, J.K.; supervision, K.C., J.-A.O. and H.-H.P.; project administration, K.C. All authors have read and agreed to the published version of the manuscript.

Funding: This research was supported by a grant from the Research Program for Agricultural Science & Technology Development (Project No. PJ015797), National Institute of Agricultural Science, Rural Development Administration, Korea. It was also partly supported by 2021 collaborative research program between the university and Rural Development Administration, Korea.

Institutional Review Board Statement: Not applicable.

Acknowledgments: The authors wish to thank the beekeepers of the Sericulture and Apiculture Division for their help.

Conflicts of Interest: The authors declare no conflict of interest. The funders had no role in the design, conduct, investigation, analysis, or writing of the study.

References

1. Potts, S.G.; Biesmeijer, J.C.; Kremen, C.; Neumann, P.; Schweiger, O.; Kunin, W.E. Global pollinator declines: Trends, impacts and drivers. *Trends Ecol. Evol.* **2010**, *25*, 345–353. [CrossRef]
2. Gallai, N.; Salles, J.M.; Settele, J.; Vaissiere, B.E. Economic valuation of the vulnerability of world agriculture confronted with pollinator decline. *Ecol. Econ.* **2009**, *68*, 810–821. [CrossRef]
3. Lebuhn, G.; Droege, S.; Connor, E.F.; Gemmill-Herren, B.; Potts, S.G.; Minckley, R.L.; Griswold, T.; Jean, R.; Kula, E.; Roubik, D.W.; et al. Detecting insect pollinator declines on regional and globalscales. *Conserv. Biol.* **2013**, *27*, 113–120. [CrossRef]
4. Neumann, P.; Carreck, N.L. Honey bee colony losses. *J. Apic. Res.* **2010**, *49*, 1–6. [CrossRef]
5. Van der Zee, R.; Pisa, L.; Andonov, S.; Brodschneider, R.; Charriere, J.D.; Chlebo, R.; Coffey, M.F.; Crailsheim, K.; Dahle, B.; Gajda, A.; et al. Managed honey bee colony losses in Canada, China, Europe, Israel and Turkey, for the winters of 2008-9 and 2009-10. *J. Apic. Res.* **2012**, *51*, 100–114. [CrossRef]
6. Vanbergen, A.J.; Baude, M.; Biesmeijer, J.C.; Britton, N.F.; Brown, M.J.F.; Brown, M.; Carvell, C. Threats to an ecosystem service: Pressures on pollinators. *Front. Ecol. Environ.* **2013**, *11*, 251–259. [CrossRef]
7. Fairbrother, A.; Purdy, J.; Anderson, T.; Fell, R. Risks of neonicotinoid insecticides to honey bees. *Environ. Toxicol. Chem.* **2014**, *33*, 719–731. [CrossRef]
8. Staveley, J.P.; Law, S.A.; Fairbrother, A.; Menzie, C.A. A causal analysis of observed declines in managed honey bees (*Apis mellifera*). *Hum. Ecol. Risk Assess.* **2014**, *20*, 566–591. [CrossRef] [PubMed]

9. VanEngelsdorp, D.; Speybroeck, N.; Evans, J.D.; Nguyen, B.K.; Mullin, C.; Frazier, M.; Frazier, J.; Cox-Foster, D.; Chen, Y.; Tarpy, D.R.; et al. Weighing risk factors associated with bee colony collapse disorder by classification and regression tree analysis. *J. Econ. Entomol.* **2010**, *103*, 1517–1523. [CrossRef]

10. Ratnieks, F.L.W.; Carreck, N.L. Clarity on Honey Bee Collapse? *Science* **2010**, *327*, 152–153. [CrossRef]

11. Grillone, G.; Laurino, D.; Manino, A.; Porporato, M. Toxicity of thiametoxam on in vitro reared honey bee brood. *Apidologie* **2017**, *48*, 635–643. [CrossRef]

12. Schmehl, D.R.; Tome, H.V.V.; Mortensen, A.N.; Martins, G.F.; Ellis, J.D. Protocol for the in vitro rearing of honey bee (*Apis mellifera* L.) workers. *J. Apic. Res.* **2016**, *55*, 113–129. [CrossRef]

13. Blacquière, T.; Smagghe, G.; van Gestel, C.A.M.; Mommaerts, V. Neonicotinoids in bees: A review on concentrations, side-effects and risk assessment. *Ecotoxicology* **2012**, *21*, 973–992. [CrossRef]

14. Krupke, C.H.; Hunt, G.J.; Eitzer, B.D.; Andino, G.; Given, K. Multiple Routes of Pesticide Exposure for Honey Bees Living Near Agricultural Fields. *PLoS ONE* **2012**, *7*, e29268. [CrossRef] [PubMed]

15. Medrzycki, P.; Giffard, H.; Aupinel, P.; Belzunces, L.P.; Chauzat, M.-P.; Claßen, C.; Colin, M.E.; Dupont, T.; Girolami, V.; Johnson, R.; et al. Standard methods for toxicology research in *Apis mellifera*. *J. Apic. Res.* **2013**, *52*, 1–60. [CrossRef]

16. Alix, A.; Chauzat, M.P.; Duchard, S.; Lewis, G.; Maus, C.; Miles, M.J.; Pilling, E.D.; Thompson, H.M.; Wallner, K. Environmental risk assessment scheme for plant protection products. Chapter 10: Honeybees-proposed scheme. *Julius-Kühn-Archiv* **2010**, *423*, 27–33.

17. Aupinel, P.; Fortini, D.; Michaud, B.; Medrzyck, P.; Padovani, E.; Przygoda, D.; Maus, C.; Charriere, J.D.; Kilchenmann, V.; Riessberger-Galle, U.; et al. Honey bee brood ring-test: Method for testing pesticide toxicity on honeybee brood in laboratory conditions. *Julius-Kühn-Archiv* **2009**, *423*, 96–101.

18. Hendriksma, H.P.; Hartel, S.; Steffan-Dewenter, I. Honey bee risk assessment: New approaches for in vitro larvae rearing and data analyses. *Methods Ecol. Evol.* **2011**, *2*, 509–517. [CrossRef]

19. Mortensen, A.N.; Ellis, J.D. A honey bee (*Apis mellifera*) colony's brood survival rate predicts its in vitro-reared brood survival rate. *Apidologie* **2018**, *49*, 573–580. [CrossRef]

20. Brødsgaard, C.J.; Ritter, W.; Hansen, H. Response of in vitro reared honey bee larvae to various doses of Paenibacillus larvae larvae spores. *Apidologie* **1998**, *29*, 569–578. [CrossRef]

21. Peng, Y.S.C.; Mussen, E.; Fong, A.; Montague, M.A.; Tyler, T. Effects of chlortetracycline on honey bee worker larvae reared in vitro. *J. Invertebr. Pathol.* **1992**, *60*, 127–133. [CrossRef]

22. Kaftanoglu, O.; Linksvayer, T.A.; Page, R.E. Rearing Honey Bees, *Apis mellifera*, in vitro 1: Effects of Sugar Concentrations on Survival and Development. *J. Insect Sci.* **2011**, *11*, 1–10. [CrossRef] [PubMed]

23. Rembold, H.; Lackner, B. Rearing of honey bee larvae in vitro: Effect of yeast extract on queen differentiation. *J. Apic. Res.* **1981**, *20*, 165–171. [CrossRef]

24. Tavares, D.A.; Roat, T.C.; Carvalho, S.M.; Silva-Zacarin, E.C.M.; Malaspina, O. In vitro effects of thiamethoxam on larvae of Africanized honey bee *Apis mellifera* (Hymenoptera: Apidae). *Chemosphere* **2015**, *135*, 370–378. [CrossRef]

25. Dai, P.; Jack, C.J.; Mortensen, A.N.; Bustamante, T.A.; Bloomquist, J.R.; Ellis, J.D. Chronic toxicity of clothianidin, imidacloprid, chlorpyrifos, and dimethoate to *Apis mellifera* L. larvae reared in vitro. *Pest Manag. Sci.* **2019**, *75*, 29–36. [CrossRef]

26. Rhein, W.V. Über die Entstehung des weiblichen Dimorphismus im Bienenstaate. Wilhelm Roux Arch. *Entwickl. Mech. Org.* **1933**, *129*, 601–665. [CrossRef] [PubMed]

27. Vandenberg, J.D.; Shimanuki, H. Technique for rearing worker honey bees in the laboratory. *J. Apic. Res.* **1987**, *26*, 90–97. [CrossRef]

28. Wittmann, D.; Engels, W. Development of test procedures for insecticide-induced brood damage in honey bees. *Mitt. Dtsch. Ges. Allg. Angew. Entomol.* **1981**, *3*, 187–190.

29. Davis, A.R.; Solomon, K.R.; Shuel, R.W. Laboratory studies of honey bee larval growth and development as affected by systemic insecticides at adult-sublethal levels. *J. Apic. Res.* **1988**, *27*, 146–161. [CrossRef]

30. OECD. *Test No. 237: Honey Bee (Apis mellifera) Larval Toxicity Test, Single Exposure. OECD Guidance for the Testing of Chemicals, Section 2, No. 237*; OECD Publishing: Paris, France, 2013.

31. OECD. *Test No. 239: Honey Bee (Apis mellifera) Larval Toxicity Test, Repeated Exposure. OECD Environment, Health and Safety Publications, Series on Testing and Assessment, No.239*; OECD Publishing: Paris, France, 2016.

32. Crailsheim, K.; Brodschneider, R.; Aupinel, P.; Behrens, D.; Gensersch, E.; Vollmann, J.; Riessberger-Gallé, U. Standard methods for artificial rearing of *Apis mellifera* larvae. *J. Apic. Res.* **2013**, *52*, 1–16. [CrossRef]

33. Fukuda, H.; Sakagami, S.F. Worker brood survival in honey bees. *Res. Popul. Ecol.* **1968**, *10*, 31–39. [CrossRef]

34. Medrzycki, P.; Sgolastra, F.; Bortolotti, L.; Bogo, G.; Tosi, S.; Padovani, E.; Porrini, C.; Sabatini, A.G. Influence of brood rearing temperature on honey bee development and susceptibility to poisoning by pesticides. *J. Apic. Res.* **2010**, *49*, 52–59. [CrossRef]

35. Williams, G.R.; Alaux, C.; Costa, C.; Csáki, T.; Doublet, V.; Eisenhardt, D.; Kuhn, R.; McMahon, D.P.; Medrzycki, P.; Murray, T.E. Standard methods for maintaining adult *Apis mellifera* in cages under in vitro laboratory conditions. *J. Apic. Res.* **2013**, *52*, 1–35. [CrossRef]

36. Staron, M.; Sabo, R.; Staroňová, D.; Sabová, L.; Abou-Shaara, H.F. The age of honey bee larvae at grafting can affect survival during larval tests. *Environ. Exp. Biol.* **2019**, *17*, 1–4.

37. Zhu, W.; Schmehl, D.R.; Mullin, C.A.; Frazier, J.L. Four common pesticides, their mixtures and a formulation solvent in the hive environment have high oral toxicity to honey bee larvae. *PLoS ONE* **2014**, *9*, e77547.

38. Aupinel, P.; Fortini, D. Improvement of artificial feeding in a standard in vitro method for rearing *Apis mellifera* larvae. *Bull. Insectol.* **2005**, *58*, 107–111.

39. Vazquez, D.E.; Farina, W.M. Differences in pre-imaginal development of the honey bee *Apis mellifera* between in vitro and in-hive contexts. *Apidologie* **2020**, *51*, 861–875. [CrossRef]

40. Riessberger-Gallé, U.; Vollmann, J.; Brodschneider1, R.; Aupinel, P.; Crailsheim, K. Improvement in the pupal development of artificially reared honeybee larvae. *Apidologie* **2008**, *39*, 595.

41. Brodschneider, R.; Riessberger-Gallé, U.; Crailsheim, K. Flight performance of artificially reared honeybees (*Apis mellifera*). *Apidologie* **2009**, *40*, 441–449. [CrossRef]

42. Fine, J.D.; Cox-Foster, D.L.; Mullin, C.A. An inert pesticide adjuvant synergizes viral pathogenicity and mortality in honey bee larvae. *Sci. Rep.* **2017**, *7*, 40499. [CrossRef]

43. Barbosa, W.F.; Tome, H.V.V.; Bernardes, R.C.; Siqueira, M.A.L.; Smagghe, G.; Guedes, R.N.C. Biopesticide-induced be-havioral and morphological alterations in the stingless bee Melipona quadrifasciata. *Environ. Toxicol. Chem.* **2015**, *34*, 2149–2158. [CrossRef] [PubMed]

44. Yang, Y.; Ma, S.; Liu, F.; Wang, Q.; Dai, P. Acute and chronic toxicity of acetamiprid, carbaryl, cypermethrin and deltamethrin to *Apis mellifera* larvae reared in vitro. *Pest Manag. Sci.* **2019**, *76*, 978–985. [CrossRef] [PubMed]

45. Seeley, T.D.; Morse, R.A. The nest of the honey bee (*Apis mellifera* L.). *Lnsectes Sociaux* **1976**, *23*, 495–512. [CrossRef]

46. Krainer, S.; Brodschneider, R.; Vollmann, J.; Crailsheim, K.; Riessberger-Gallé, U. Effect of hydroxymethylfurfural (HMF) on mortality of artificially reared honey bee larvae (*Apis mellifera* carnica). *Ecotoxicology* **2016**, *25*, 320–328. [CrossRef] [PubMed]

47. Tehel, A.; Vu, Q.; Bigot, D.; Gogol-Döring, A.; Koch, P.; Jenkins, C.; Doublet, V.; Theodorou, P.; Paxton, R. The two prevalent genotypes of an emerging infectious disease, deformed wing virus, cause equally low pupal mortality and equally high wing deformities in host honey bees. *Viruses* **2019**, *11*, 114. [CrossRef]

48. Oertel, E. Metamorphosis in the honeybee. *J. Morphol.* **1930**, *50*, 295–339. [CrossRef]

 *insects*

Article

Characterization of Thermal and Time Exposure to Improve Artificial Diet for Western Corn Rootworm Larvae

Man P. Huynh [1,2,*], Adriano E. Pereira [1], Ryan W. Geisert [1], Michael G. Vella [3], Thomas A. Coudron [4], Kent S. Shelby [4] and Bruce E. Hibbard [5]

1   Division of Plant Science & Technology, University of Missouri, Columbia, MO 65211, USA; pereiraa@missouri.edu (A.E.P.); geisertblue@gmail.com (R.W.G.)
2   Department of Plant Protection, Can Tho University, Can Tho 900000, Vietnam
3   Frontier Scientific Services, Newark, DE 19711, USA; mvella@fsiag.com
4   Biological Control of Insects Research Laboratory, USDA-Agricultural Research Service, Columbia, MO 65203, USA; coudront@missouri.edu (T.A.C.); kent.shelby@usda.gov (K.S.S.)
5   Plant Genetics Research Unit, USDA-Agricultural Research Service, Columbia, MO 65211, USA; bruce.hibbard@usda.gov
*   Correspondence: manhuynh@missouri.edu

Citation: Huynh, M.P.; Pereira, A.E.; Geisert, R.W.; Vella, M.G.; Coudron, T.A.; Shelby, K.S.; Hibbard, B.E. Characterization of Thermal and Time Exposure to Improve Artificial Diet for Western Corn Rootworm Larvae. *Insects* **2021**, *12*, 783. https://doi.org/10.3390/insects12090783

Academic Editor: Allen Carson Cohen

Received: 16 August 2021
Accepted: 29 August 2021
Published: 1 September 2021

Publisher's Note: MDPI stays neutral with regard to jurisdictional claims in published maps and institutional affiliations.

Simple Summary: The western corn rootworm is a highly adaptive pest that has evaded nearly all management tactics developed to date. Antibiotics have been utilized in rootworm diets to mitigate bacterial contamination. However, antibiotic ingestion necessarily alters rootworm gut microbiota, clouding the outcome of diet toxicity bioassays used in determination of rootworm susceptibility to insecticides. Rapid heating, or pasteurization, is one of the most widely applied techniques to alleviate microbial contamination and could eliminate antibiotics from the diet. We characterized effects of temperatures and time intervals of thermal exposure on quality of rootworm diet by measuring larval weight, molting, and survival. Our results demonstrated non-linear effects of thermal exposure on the performance of diet, whereas no impacts were observed on the exposure intervals evaluated. These findings will guide the continued development of sterilized rootworm diets, facilitating mass production and provide insights into the design of diets for other insects.

Abstract: The western corn rootworm (WCR), *Diabrotica virgifera* LeConte, is the most serious pest of maize in the United States. In pursuit of developing a diet free of antibiotics for WCR, we characterized effects of thermal exposure (50–141 °C) and length of exposure on quality of WCRMO-2 diet measured by life history parameters of larvae (weight, molting, and survival) reared on WCRMO-2 diet. Our results indicated that temperatures had non-linear effects on performance of WCRMO-2 diet, and no impacts were observed on the length of time exposure. The optimum temperature of diet processing was 60 °C for a duration less than 30 min. A significant decline in development was observed in larvae reared on WCRMO-2 diet pretreated above 75 °C. Exposing WCRMO-2 diet to high temperatures (110–141 °C) even if constrained for brief duration (0.9–2.3 s) caused 2-fold reduction in larval weight and significant delays in larval molting but no difference in survival for 10 days compared with the control diet prepared at 65 °C for 10 min. These findings provide insights into the effects of thermal exposure in insect diet processing.

Keywords: *Diabrotica virgifera*; corn rootworm; WCRMO-2; diet processing; heating

1. Introduction

The western corn rootworm (WCR), *Diabrotica virgifera virgifera* LeConte (Coleoptera: Chrysomelidae), is the most serious pest of maize in the United States and some parts of Europe [1], causing 1 to 2 billion dollars (USD) in losses and control costs to U.S. maize growers each year [2]. Most damage associated with this species is the result of larval feeding on maize roots [3,4], though yield reduction of maize can result from adult

Insects **2021**, *12*, 783. https://doi.org/10.3390/insects12090783   https://www.mdpi.com/journal/insects

feeding on silks, pollen, kernels, and foliage of maize plants [5]. Management of WCR has been a challenge because this highly adaptive insect has evolved resistance to several management strategies, including chemical insecticides [6–9], transgenic maize hybrids expressing insecticidal crystalline toxins from *Bacillus thuringiensis* (Bt) Berliner [10–15], and cultural control techniques, such as crop rotation [16,17].

Given a history of developing resistance to nearly every management tactic utilized for managing WCR, a logical concern exists that WCR will possibly develop resistance to newer management tactics. To slow resistance development of this pest, the U.S. Environmental Protection Agency (EPA) has mandated monitoring resistance programs that involve annual collections of insect populations in regions of high adoption of the targeted trait followed by bioassays to determine potential reduction in susceptibility attributable to resistance development [18]. Diet assays, whereby insects are exposed to toxins in an artificial diet, that can be used in conjunction with on-plant assays to evaluate the susceptibility of WCR to insecticides are critical components of the resistance-monitoring programs [19–21].

An artificial diet capable of supporting WCR larval growth and development similar to those fed on maize roots would be greatly beneficial for research programs. Artificial diet development for *Diabrotica* spp. was initially conducted on the southern corn rootworm, *Diabrotica undecimpunctata howardi* Barber, as a model species [22,23] because diet work began prior to availability of the non-diapausing strain of WCR. An attempt to develop a diet for WCR rearing occurred in 2002 [24]. Later, Huynh et al. [25] developed an improved WCR diet (WCRMO-1) that was an optimization of the ingredients in the initial WCR diet [24]. The WCRMO-1 formulation is compatible with each of the four marketed Bt toxins targeting WCR [26]. However, both published WCR diets [24,25] require maize root powder, which is not available for purchase, thereby limiting the practical use of the diets. Recently, Huynh et al. [27] successfully developed a WCR diet without maize root powder (WCRMO-2) that supports performance of WCR larvae equal to or better than that of publicly available formulations [27,28]. The WCRMO-2 formulation was specifically designed to require only commercially available ingredients. This formulation supported approximately 97% of larvae for survival, molting, and increased larval weight gain after 10 days of feeding by 4-fold compared with the WCRMO-1 diet [27]. Both WCRMO-1 and WCRMO-2 diets have essentially zero microbial contamination [27], similar to the four proprietary diets previously used for WCR bioassays [26,28], through clean laboratory practices described previously [25]. The WCRMO-2 diet is now available commercially (WCRMO-2, Frontier Scientific Services, Newark, DE, USA).

Antibiotics are commonly used in insect diets for preventing bacterial contamination. Ingestion of antibiotics in insect diets has been reported to alter response of lepidopteran insects to Bt proteins due to its negative effects on insect gut microbiomes [29,30]. In fact, Paramasiva, Sharma, and Krishnayya [29] manipulated antibiotics to suppress gut microflora in the cotton bollworm (*Helicoverpa armigera* Hübner) and found that *H. armigera* larvae fed on larval diet with antibiotics had lower susceptibility to a commercial formulation of Bt and purified δ-endotoxins Cry1Ab and Cry1Ac compared with *H. armigera* larvae fed on larval diet without antibiotics. Additionally, pretreatment of *H. armigera* larvae with antibiotics to eliminate the gut microbes resulted in a decrease in larval mortality and an increase in the larval weight gain when *H. armigera* larvae were exposed to activated Cry1Ac, Bt formulation, and transgenic cotton [30]. In WCR, antibiotics have been reported to cause effects on the indigenous symbionts that contribute the development of resistance in WCR to crop rotation [31]. Differences in the abundance of multiple bacterial taxa (e.g., *Acinetobacter* sp., *Pseudomonas* sp., *Enterobacter* sp., *Lactococcus* sp.) in WCR gut microbiota were correlated with increased resistance to soybean-defense compounds. Gut microbiota of WCR derived from rotation-resistant populations and wild-type populations had differences in the abundance of *Klebsiella* sp., *Stenotrophomonas* sp., *Enterobacter* sp., and *Lactococcus* sp. [31]. The authors suppressed gut microbiota of rotation-resistant WCR by antibiotic treatments, finding that the resistance to the soybean-defensive compound of the resulting rotation-resistant WCR was reduced to a level similar to that of wild-type

WCR. Paddock et al. [32] reported that feeding on maize expressing a Bt toxin resulted in a shift in the gut microbiota in susceptible WCR larvae but led to no changes in the bacterial community within resistant insects. Currently, published diet formulations for *Diabrotica* spp. contain antibiotics (streptomycin and chlortetracycline) [22,24,27,33] that are widely used in other insect diets as the important antibacterial agents [34]. It is likely that these antibiotics have potential impacts on WCR gut microbiomes, though these effects of antibiotics on WCR larvae are not adequately characterized. The use of a diet free from antibiotics for WCR bioassays would likely provide more accurate phenotypic picture of test populations, as it relates to susceptibility to Bt toxins and other insecticide compounds. Therefore, development of a WCR diet free of antibiotics would facilitate resistance-monitoring programs as well as other research programs of this pest.

One of the most widely used techniques to alleviate microbial contamination in insect diets is thermal treatment as a means of diet preservation [34]. During diet processing, agar-based diets are often exposed to elevated temperatures ranging from mild blanching (50–100 °C) to high temperatures above the boiling point of water. These high-thermal treatments are typically manipulated with pressure, which can lead to the destruction of microbial contaminants. Flash sterilization is one of the heating techniques that utilizes high temperatures for a brief time of thermal exposure (less than 1 min) that sufficiently kills microbial contaminants in the diet [34] while minimizing the effects of overheating on heat-sensitive nutrients (e.g., vitamins, amino acids, or lipids) [35]. As a step toward development of a diet free of antibiotics for WCR, we investigated the effects of high temperatures (80–141 °C) for short time exposure (0.8–2.3 s) on quality of WCRMO-2 diet using a flash sterilization approach. Additionally, since the effects of using relatively low temperatures on the performance of insect diets have not been adequately explored, we further characterized the effects the mild thermal exposure (50–80 °C) and extended time of thermal exposure (5–30 min), which are commonly used in insect diet preparation, on the quality of WCRMO-2 diet. The quality of WCRMO-2 diet was assessed via the evaluation of life history parameters (weight, molt, and survival) of WCR larvae reared on the diets for 10 days.

2. Materials and Methods

2.1. Insects and Egg Treatment

Eggs of non-diapausing WCR populations were obtained from the USDA-ARS-Plant Genetics Research Unit (PGRU) laboratory in Columbia, MO. Egg plates containing WCR eggs and soil in Petri dishes were incubated at 25 °C in complete darkness until ~5% of eggs were hatched. Subsequently, the eggs were washed from soil with tap water and then were surface-sterilized using undiluted Lysol (Clean & Fresh Multi-Surface Cleaner, Reckitt Benckiser, Parsippany, NJ, USA) for 3 min and followed by 10% formaldehyde (HT501128, Sigma Aldrich, St. Louis, MO, USA) for 3 min, as described previously [24,36]. The eggs were dispensed onto coffee filter paper (Pure Brew, Rockline Industries, Sheboygan, WI, USA) placed inside a 16 oz. cup (11.7 × 7.62 × 9.6 cm, DM16R-0090, Solo Cup Company, Lake Forest, IL, USA) with a lid (LG8RB-0090, Solo Cup Company) using a 1-mL disposable pipette (13-711-9a, Fisher Scientific, Pittsburg, PA, USA). The eggs were then incubated at 25 °C in darkness. Larvae that hatched within 24 h were used for insect bioassay.

2.2. Experimental Approach

A series of experiments were performed to determine the effects of thermal and time exposure on the quality of WCRMO-2 diet. WCRMO-2 diet was made using WCRMO-2 dry mix (WCRMO-2, Frontier Scientific Services) at different temperatures and varying lengths of time of exposure to the different temperatures. The WCRMO-2 dry mix (Frontier Scientific Services) consists of ingredients, diet preservatives (antifungal and antibacterial agents), and agar that were published previously [27]. The quality of WCRMO-2 diet was evaluated by life history parameters of WCR larvae (weight, molt, and survival) reared on the diet treatments for 10 days.

Two-level factorial designs were constructed to determine the effects of two factors (temperature and time of thermal exposure) on the quality of WCRMO-2 diet. Two experiments were performed to evaluate high temperatures and brief time of thermal exposure (80–141 °C and 0.8–2.3 s) and mild temperatures and extended time of thermal exposure (50–80 °C and 5–30 min). The high temperature and the brief time of thermal exposure (141 °C for 2 s) was previously used for flash sterilization of a larval diet for *Trichoplusia ni* (Hubner) (Lepidoptera: Noctuidae) [37]. Since preliminary observations indicated that time of thermal exposure up to 10 min at 65 °C had no effects on the quality of WCRMO-2 diet, the extended time of thermal exposure up to 30 min was selected to further determine the effects of the longer duration of thermal exposure. The experimental designs for each experiment were generated with Design-Expert (Stat-Ease, Inc., Minneapolis, MN, USA). All designs consisted of 11 design points (diet treatments at different thermal exposure and time of exposure) with 5 model, 3 lack of fit, and 2 pure error degrees of freedom. In all experiments, WCRMO-2 diet made using the standard protocol according to the manufacturer's procedure, with a temperature of 65 °C and time of exposure for 10 min [27], was included as a control (Table 1).

Table 1. Diet treatments used in 2-factorial experiments to rear western corn rootworm larvae.

Treatment	High Temperature		Mild Temperature	
	Temperature (°C)	Time Exposure (Second)	Temperature (°C)	Time Exposure (Minute)
1	141	1.47	80	17
2	132	1.07	76	26
3	132	2.00	76	10
4	110	0.80	65	30
5	110	1.47	65	17
6	110	1.47	65	17
7	110	1.47	65	17
8	110	2.27	65	5
9	88	1.07	55	26
10	88	2.00	55	9
11	80	1.47	50	17
12 (control)	65	600	65	10

A flash sterilization system (Frontier Scientific Services) was utilized to produce WCRMO-2 diet at high thermal exposure (80–141 °C) and short time exposure (0.8–2.3 s). We observed consistent failures in the ability of the powdered agar to melt into solution during the brief time and rather high thermal exposure experienced in treatments with temperatures less than 88 °C (Table 1). This is likely due to the physical limitations of the agar (7060, Frontier Scientific Services) used to operate adequately at the mild temperatures tested. This resulted in a "soft-set" of the media, providing an inadequate matrix to support larval growth, leading to the death of the majority of larvae and increased bacterial contamination after 4 days post infestation. Low-melt agar (Frontier Scientific Services) was tested as a substitution of the agar used initially. However, the "soft-set" of the media was observed when the media were prepared with the low melt agar at the temperatures $\leq$ 65 °C for small windows of time (0.9–2.1 s). To further study the effects of mild thermal exposure (50–80 °C) on the quality of WCRMO-2 diet, an alternative approach involved completely melting the agar using a microwave (51101BZ, Hamilton Beach, Glen Allen, VA, USA) and cooling it to designed mild temperatures prior to adding it to the treated media was evaluated. The temperatures were then held for designed time exposure (5–30 min) using a hot plate (Cimarec^{TM}, Thermo Scientific, Waltham, MA, USA) (Table 1).

3. Diet Preparation

At high thermal exposure (80–141 °C), WCRMO-2 diet was made using a flash sterilization system (Frontier Scientific Services). Frontier's sterilizer consists of a circulating system of processing tubes that alternates between linear stretches of tubing and coiled spiraltherms that act as heat exchangers (Figure 1). To make 10 L of WCRMO-2 diet, 1.49 Kg of WCRMO-2 dry mix (WCRMO-2, Frontier Scientific Services) and 158 g of agar (7060, Frontier Scientific Services) were added to 9.26 L of cool tap water (~21 °C). The diet mixture was then pumped into the process lines from a tank and controlled via a metering pump that manages line pressure and flowrate of the product. The time of thermal exposure was set by manipulating the flowrate. The first heat exchanger was jacketed with hot oil set to a desired temperature. As the product flows through the coiled spiraltherm tubing, it rapidly gains thermal units from the hot oil jacketing the line. After departing the heated coil, the process line enters a series of linear switch-backs so that some excess heat is lost to atmosphere before reaching the cooling spiraltherm. At this point, the product enters a heat exchanger jacketed in cold (room temperature) water to rapidly lose thermal units and reach a desired dispense temperature. At the dispense temperature of 60 °C, 289.5 mL of 10% KOH (*w*/*v*) (F7633, Frontier Scientific Services) was added into the diet mixture to adjust the pH of the diet to 9, and the resulting diet solution was blended for 30 s to mix thoroughly. Subsequently, the diet solution was dispensed into a 96-well plate (3370, Corning Inc., Corning, NY, USA) using a repeater pipette (200 μL per well) in a biological cabinet (Forma 1800 Series Clean Bench, Thermo Scientific). The diet plate was then allowed to evaporate excess moisture for 10 min, stored in a refrigerator at 7 °C, and used for assays within 3 days.

Figure 1. Frontier Scientific's flash sterilization system used for the processing of insect media. From left to right: diet tank, pump, processing tubes, coiled spiraltherms, and control station.

At mild thermal exposure (50–80 °C), WCRMO-2 diet was made using WCRMO-2 dry mix (WCRMO-2, Frontier Scientific Services) according to the manufacturer's procedure [27]. To prepare 1 L of WCRMO-2 diet, agar (15.8 g) was added to 926 mL of purified water, and the solution was brought to a full boil using a microwave (Hamilton Beach) until agar was completely melted. The agar solution was then transferred to a blender placed in a biological safety cabinet (SG403, SterilGARD® III Advance cabinet, Sanford, ME, USA). When the agar solution cooled to designed temperatures (Table 1), 148.9 g of WCRMO-2 dry mix was added, and the mixture was blended for 10 s to mix thoroughly. Subsequently, 28.95 mL of 10% KOH (*w*/*v*) (P250, Fisher Scientific, Fair Lawn, NJ, USA) was added to increase the pH of the diet to 9, and the resulting diet solution was blended for 10 s to mix thoroughly. The diet was poured into a 1-L glass beaker containing a stir-bar and placed on a stirring hot plate (Cimarec^{TM}, Thermo Scientific) set at the designed temperatures. The temperatures of diet solution were monitored using an infrared thermometer (IR002, Ryobi, Fuchu, Hiroshima, Japan) and held at the test temperatures for the designed times (5–30 min) using the hot plate. Subsequently, the diet solution was dispensed into a 96-well plate (3370, Corning Inc.) using a repeater pipette (200 µL per well), allowed to evaporate excess moisture for 10 min, stored in a refrigerator at 4 °C, and used for assays within 3 days.

3.1. Insect Artificial Diet Bioassays

The diet bioassays were conducted as described previously [25]. All materials used in the diet assays were surface-treated via exposure to UV light for 10 min in a biological cabinet (SterilGARD® III Advance cabinet). Each diet treatment, which is WCRMO-2 diet made at different temperatures and time of thermal exposure (Table 1), was randomly assigned to a 12-well row of the 96-well plate and replicated at least 3 times in different diet plates. Each well was infested with one WCR neonate (<24 h old) using a fine paintbrush. A sealing film (TSS-RTQ-100, Excel Scientific, Inc., Victorville, CA, USA) was used to cover the plate. For ventilation, a hole was made in the sealing film over each well using a number zero insect pin. The plates were kept in an incubator (Percival, Perry, IA, USA) at 25 °C in darkness for 10 days. Larval molting was recorded daily during the experiments, whereas larval weight, survival, and evidence of contamination were recorded at the end of the experiments. For larval dry weight, all live larvae in each treatment were pooled per replicate (12 possible) into 95% ethanol, dried in an oven (Binder GmbH, Tuttlingen, Germany) at 55 °C for 48 h, and weighed using a micro balance (MSU6.6S-000-DM, Sartorius Lab Instruments GmbH & Co. KG, Goettingen, Germany).

3.2. Data Processing and Statistical Analyses

Survival and molting data were calculated by dividing the number of live larvae and successful larval molt from 1st to 2nd instar and from 1st to 3rd instar per replicate, respectively, by the initial number of larvae infested and multiplying by 100 to obtain percentages. Weight per larva (mg) was determined by dividing the dry weight by the number of larvae that survived per replicate.

For the experiment with high temperatures (80–141 °C) and short time of thermal exposure, the diet treatments that resulted in a "soft-set" of the media were excluded in the analyses because the soft-set form led to the death of the majority of larvae after 4 days post infestation. Because of the exclusion of these treatments, the remaining data points were not adequate to generate the response surface models of the measured responses. The remaining data were analyzed as a completely randomized experiment. The data were analyzed with analysis of variance (ANOVA) using PROC MIXED in SAS 9.4 (SAS Institute, Cary, NC, USA). Diet was the fixed effect, and replication was the random variable. Differences between the remaining treatments were determined using Fisher's least significant difference (LSD) at $p < 0.05$. The percent variables (survival and molting) were arcsine square-root transformed prior to the analysis to meet assumptions of normality and homoscedasticity, whereas untransformed data were presented as mean $\pm$ SEM.

For the experiment with mild temperatures (50–80 °C) and extended time of thermal exposure, polynomial equations were generated to describe the impact of two factors (temperature and time of thermal exposure) on the measured responses (larval weight, molting, and survival). The best fit model for each measured response was selected from all possible models from linear to quartic polynomials generated with Design Expert® (Stat-Ease, Inc., Minneapolis, MN, USA). Model selection was based on several criteria, including low model p-value, lack of fit p-value, low standard deviation, high R-values, and a low PRESS value [38,39]. Once more than one satisfactory model was generated, adequacy tests were performed to further evaluate the selected model, as described previously [40].

4. Results

4.1. Diet Quality with High Thermal Exposure and Short Time of Thermal Exposure

Exposure to high temperatures (110–141 °C), even if constrained to a small window of time (0.9–2.3 s), had significant deleterious effects on the quality of WCRMO-2 diet for feeding WCR larvae. The WCRMO-2 diet exposed to the high temperatures for the brief time exposure resulted in larval weight significantly smaller compared to the control diet, WCRMO-2 diet made at a temperature of 65 °C for 10 min ($p < 0.0001$, $F_{6,13} = 15.88$. Figure 2a). Average larval dry weight on the diet treatments ranged from 0.30 mg to 0.39 mg, while average dry weight on the control diet was 0.71 mg, an approximately 2-fold difference. There was no significant difference in dry weight on all diet treatments. Significant delays in larval molt to the 2nd instar were observed when they were reared on all diet treatments compared with the control diet (Figure 2b). There were significantly fewer 2nd instar larvae on the diet treatments than on the control diet after 5 days post infestation when larvae began to molt to 2nd instar. At day 5 post infestation, 20.6% of larvae had molted to 2nd instar on the control diet, whereas nearly 0% of 2nd instar larvae on all diet treatments ($p < 0.0001$, $F_{6,32} = 37.60$). At day 7 post infestation, nearly 100% of the larvae had molted to 2nd instar on the control diet, significantly higher than that of all diet treatments ($p < 0.0001$, $F_{6,32} = 19.82$). There was no significant difference in percent larvae molted to 2nd instar in the diet treatments (range from 39.3%–48.4%) at day 7 post infestation. No larvae that had molted to the 3rd instar were observed on the diet treatments by 10 days. Larval survivorship on all diet treatments ranged from 95.7% to 100%, which was not significantly different from survivorship on the control diet ($p = 0.7764$, $F_{6,13} = 0.53$, Figure 2c).

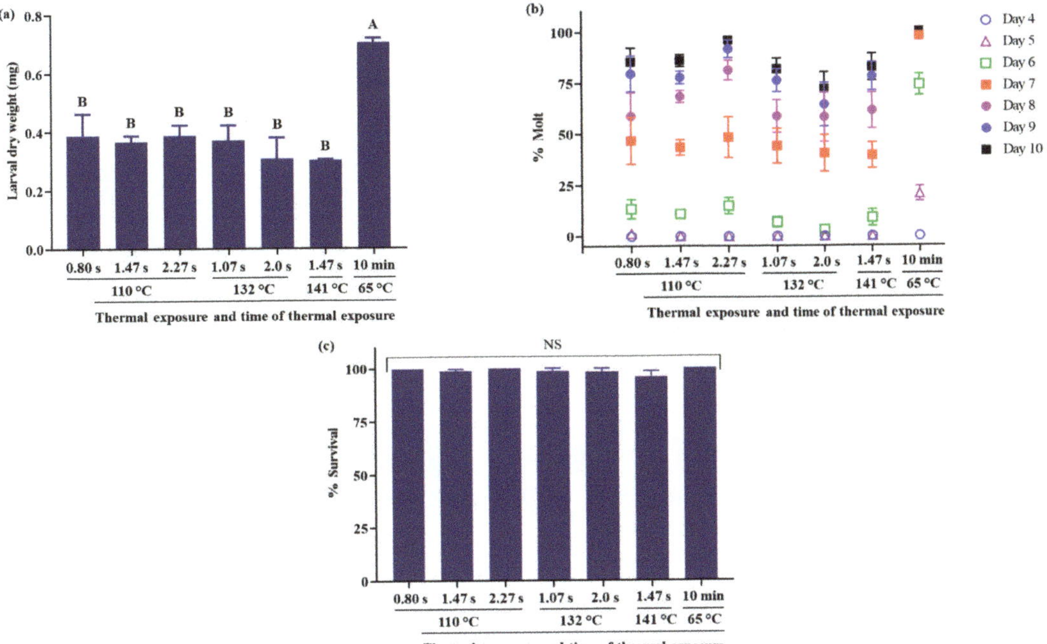

Figure 2. Effects of high thermal exposure (110–141 °C) and short time intervals of thermal exposure (0.8–2.3 s): dry weight (**a**), percent successful completion of molt to 2nd instar (**b**), and percent survival (**c**). Western corn rootworm larvae were reared on WCRMO-2 diet for 10 days. Means with bars followed by different letters are significantly different ($p < 0.05$). Means ± SEM.

4.2. Diet Quality with Mild Thermal Exposure and Extended Time of Thermal Exposure

The two-level factorial experiment yielded significant response surface models for larval weight ($p < 0.0015$, $F_{3,7} = 16.27$), molt to 2nd instar ($p = 0.0410$, $F_{3,7} = 4.76$), molt to 3rd instar ($p = 0.0100$, $F_{3,7} = 8.46$) and a marginally significant model for survival ($p = 0.0531$, $F_{3,7} = 4.23$) (Table 2). Models for weight, molt to 3rd instar, and survival had insignificant lack of fit, whereas there was a significant lack of fit for molt to 2nd instar due to a very small value of pure error. The relationships between the two factors tested (mild thermal exposure and extended time of exposure) and the quality of WCRMO-2 diet were revealed in contour plots (Figure 3). The contour plots generated from the models of responses measured (weight, survival, and molting) displayed the performance of larvae when reared on the WCRMO-2 diet prepared at the mild temperatures for the extended duration. The magnitude of the response variables is coded in color and can be envisioned as perpendicular to the page, as indicated by labelled isobars.

Table 2. *p*-values, regression coefficients, and response surface-model fitting diagnostic statistics for western corn rootworm responses to a 2-factorial experiment at mild thermal exposure (50–80 °C) and extended time of thermal exposure (5–30 min). A: time of thermal exposure, B: temperature.

	Weight *p*-Values	Regression Coefficients	% Molt to 2nd Instar *p*-Values	Regression Coefficients	% Molt to 3rd Instar *p*-Values	Regression Coefficients	Survival *p*-Values	Regression Coefficients
Model	0.0015	-	0.0410	-	0.0100	-	0.0531	-
A	0.8538	−0.0005	0.8940	0.0005	0.5604	0.0098	0.0582	−0.0088
B	0.0012	0.1321	0.0224	0.0969	0.0051	0.0511	0.1563	−0.0033
AB	-	-	-	-	-	-	0.0596	0.0002
B²	0.0025	−0.0011	0.0478	−0.0008	0.0204	−0.0004	-	-
Lack of fit	0.1953		0.0005		0.9195		0.5496	
Model type	Quadratic (reduced)		Quadratic (reduced)		Quadratic (reduced)		Two-factor interaction	
R²	0.8745		0.6710		0.7838		0.6443	
R²_adj	0.8208		0.5300		0.6911		0.4919	

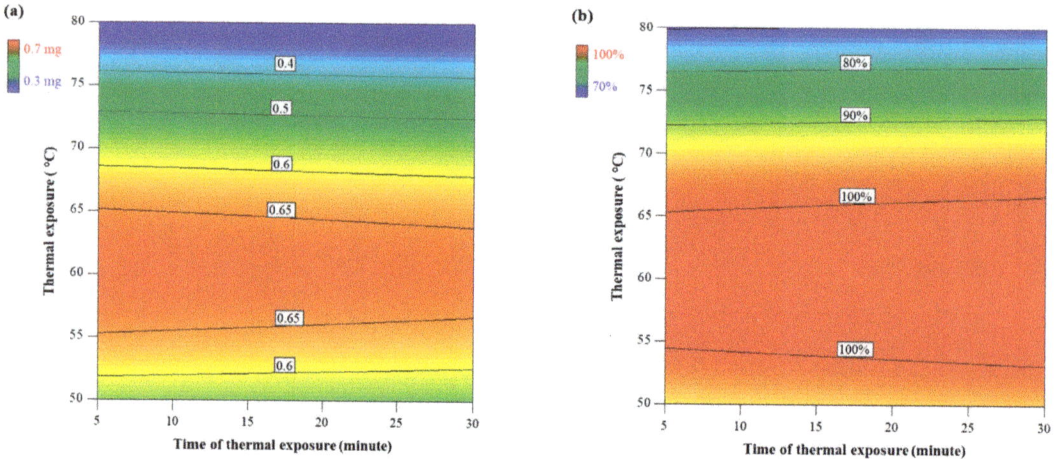

Figure 3. *Cont.*

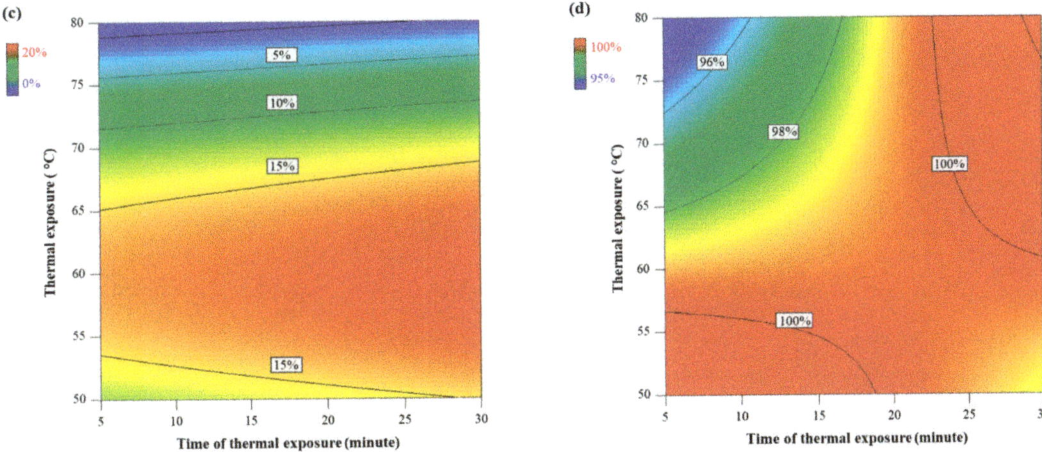

Figure 3. Effects of mild thermal exposure (50–80 °C) for extended times of thermal exposure (5–30 min). Contour plots of dry weight (**a**), percent successful completion of molt to 2nd instar (**b**) and percent successful completion of molt to 3rd instar (**c**), and percent survival (**d**). Western corn rootworm larvae were reared on WCRMO-2 diet for 10 days. Color bars display the magnitude of the measured responses.

Models for weight, molt to 2nd instar, and molt to 3rd instar revealed that temperature had significant effects on these measured responses, while no significant effect on weight and molt was observed due to time of thermal exposure. p values of temperature were 0.0012, 0.0224, and 0.0051 in the models for weight, molt to 2nd instar, and molt to 3rd instar, respectively (Table 2). Non-linear effects of temperature on weight, molt to 2nd instar, and molt to 3rd instar were found (Figure 4). WCRMO-2 diet prepared a temperature of approximately 60 °C yielded the maximum larval weight and percent molt, whereas there were significantly negative effects on weight and molting when WCRMO-2 diet was produced at temperatures above 75 °C (Figures 3a–c and 4).

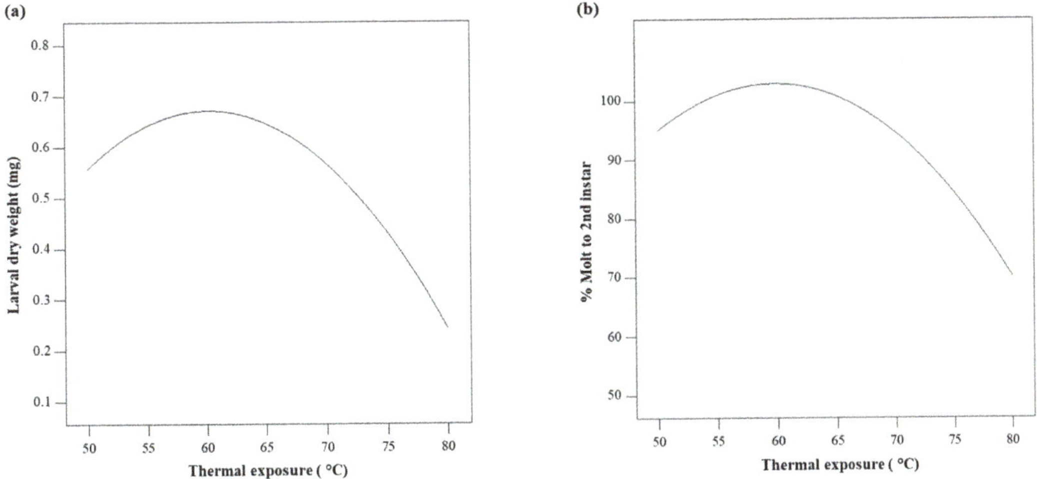

Figure 4. *Cont.*

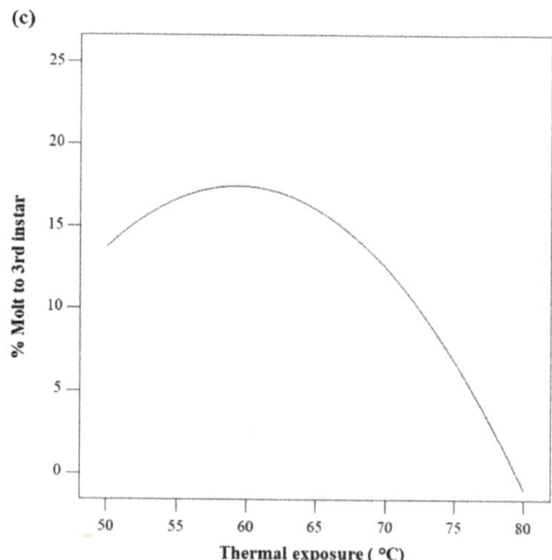

Figure 4. Nonlinear effects of mild thermal exposure (50–80 °C) for 15 min on dry weight (**a**), percent successful completion of molt to 2nd instar (**b**), and percent successful completion of molt to 3rd instar (**c**). Western corn rootworm larvae reared on WCRMO-2 diet for 10 days.

A model for survival indicated marginally significant effects of time exposure ($p = 0.0582$) and the interaction between thermal exposure and time exposure ($p = 0.0596$). However, larval survivorship on all diet treatments was >95% (Figure 3d). A similar pattern was found in the experiment with high thermal exposure and short time of thermal exposure (Figure 2c). Consequently, survival was not considered as an important criterion for evaluation of the effects of the treatments.

4.3. Contamination

All experiments had minor contamination (<3%) except for diet treatments that experienced a soft-set of the media that were excluded from the analyses. No evidence for a relationship between contamination and two experimental factors (thermal exposure and time of thermal exposure) was determined. Similarly low contamination rates were observed previously [27].

5. Discussion

Published WCR diets require antibiotics as diet preservatives to alleviate bacterial contamination [24,25,27]. However, ingestion of antibiotics results in changes in WCR gut microbiota [31], thereby possibly interfering with the determination of the susceptibility of WCR to insecticide toxins using diet bioassays. The availability of a WCR diet free of antibiotics would facilitate research programs of this important pest. In diet processing, heating is required as one of the most important parts for activating gelling agents (e.g., agar) to stabilize diets and promote the form of suitable textures for insect feeding. This can be also used as an extremely effective means for destroying microbial contaminants derived from diet ingredients and preparation [34]. To further the goal of developing a diet free of antibiotics for WCR, we explored the effects of the thermal exposure from 50–141 °C for the short time (<3 s) and extended time (10–30 min) on the quality of WCRMO-2 diet based on life history parameters of WCR larvae fed on the treated diets. By using geometric and mathematical approaches, we further characterized the influence of both thermal exposure and time of thermal exposure, allowing determination of the optimum conditions (temperature and time) for making WCRMO-2 diet.

Our results indicated that the exposure to the high thermal exposure (110–141 °C) for brief intervals (0.9–2.3 s) caused detrimental effects on the performance of larvae on

WCRMO-2 diet, indicating a possible reduction in nutrients required for WCR growth and development due to the high thermal exposure even if constrained for the short duration. WCR larvae fed on the treated diets exhibited significant reductions in weight and molting compared to those fed on the control diet made at the mild temperature of 65 °C for 10 min, whereas no difference in larval survivorship was observed between the diet treatments and the control diet. Heating is known to provide many benefits, including destruction of microbial contaminants, activation of gelling agents, increasing protein digestibility, denaturation of digestive inhibitors and harmful enzymes (e.g., phenol oxidases, lipo-oxygenase), increasing flavor, and acceleration of desirable chemical reactions [34]. However, severe overheating can cause reduction of protein digestibility, nutrient destruction (e.g., ascorbic acid, unsaturated lipid), destruction of vitamins, and formation of complexes (sugar amino acid products) [41]. A similar pattern of negative effects of heating on the quality of insect diets was previously reported for a diet of *Chrysoperla carnea* Stephens (Neuroptera: Chrysopidae) [42]. Initially, Vanderzant [43] developed a liquid diet for *C. canea* larvae that contains soy and casein hydrolysates as the main protein sources along with fructose, vitamins (ascorbic acid, B-vitamins), and other diet ingredients. This formulation was successfully used for the production of 18 generations of *C. canea*, but it did not promote the growth of *C. canea* after autoclaving at 121 °C [42]. The author successfully developed an improved diet that can be autoclaved by removing ascorbic acid (a heat-sensitive ingredient), replacing fructose with sucrose, and adding yeast hydrolysate and casein. These substitutions effectively compensated the loss of heat-sensitive nutrients. Only minor difference in percent adult recovery from larvae between the improved diet with and without autoclaving, which was 39% and 34%, respectively, was observed [42]. However, there was approximately 2-fold reduction in the percentage of adult recovery from larvae when *C. canea* larvae were reared on the standard diet [43] compared with the heat sterilized diet [42]. Griffin et al. [44] compared percent adult recovery from larvae of the boll weevils, *Anthonomus grandis* Boheman (Coleoptera: Curculionidae), reared on artificial diets that were made high temperatures of 130 °C, 138 °C, 144 °C, and 151 °C for 30 s using a flash-type sterilizer. They found that no significant effects of these temperatures on the percentage of adult recovery from larvae, though the highest temperature of 151 °C yielded fewer and smaller weevils than other temperatures. Our ongoing efforts to develop a heat-sterilized diet for WCR facilitating fewer or no antibiotics focus on the identification of alternative ingredients (e.g., yeast hydrolysate, casein) that can be compensated for the loss of nutrition in WCRMO-2 diet due to the overheating.

With diet ingredients in WCRMO-2 formulation, we demonstrated that a temperature of 60 °C yielded the highest larval performance on WCRMO-2 diet, while a time exposure of less than 30 min did not have significant impacts on the quality of WCRMO-2 diet. The exposure of WCRMO-2 diet to temperatures below 55 °C or above 65 °C resulted in a reduction in larval weight and molting compared to WCRMO-2 diet made at the optimum temperature. The significant losses in nutrients needed for WCR larval growth and development were observed when the WCRMO-2 diet was heated over 75 °C. Although most insect diets are often made at temperatures of 65–70 °C [24] that are likely used to avoid the loss of nutrition of heat-sensitive ingredients (e.g., ascorbic acid, vitamins, lipids), no information on the effects of relative low temperatures (50–80 °C) on insect diets is available. This study adds to the limited number of studies characterizing the effects of thermal exposure and time of thermal exposure, especially at mild temperatures, on insect diets. Some insect diets that have been reported to be heat-tolerant are usually produced at temperatures of 121 °C for 15–20 min by autoclaving or 141 °C for a few seconds by flash sterilizing [37,44]. It is noteworthy that WCR is nearly monophagous on maize roots and can survive on a few grass species [45]. This pest may require specific nutrients, and their bioavailability in WCRMO-2 diet is significantly reduced when the diet is exposed to temperatures over 75 °C.

Diet assays determining susceptibility of corn rootworm larvae typically utilizes a 96-well microtiter plate format [46–48]. Typically, each well is filled with 200 µL of diet and overlaid with toxins and followed by an infestation of a single neonate larva and sealing the well. This assay involves a labor-intensive filling process. A high-throughput system designed to run and analyze assays on a large scale, factorially increasing the number of compounds screened in the case of discovery, or the number of assays completed in the case of insect resistance management (IRM) studies would greatly facilitate research programs of this important pest. Research with lepidopteran insects has shown the great value of high-throughput systems for mass production and diet bioassays. In fact, utilizing a flash sterilizer coupled to a form-fill seal machine allowed the mass production of 3 million corn earworm pupae, *Helicoverpa zea* (Boddie) (Lepidoptera: Noctuidae), annually [49]. More recently, the high-throughput system can produce 25,000 rearing units per hour containing artificial diet and eggs of *Trichoplusia ni* (Hubner) (Lepidoptera: Noctuidae) through the use of automation [37]. For these systems, in addition to a flash sterilizer and a form-fill seal machine, a heat-sterilized diet is one of the key components. WCR has proven to be one of the most challenging pests in North America [50]. Many management strategies have been developed for controlling WCR (crop rotation, soil insecticides, Bt maize), but this pest has evaded nearly all management tactics in recent years [10–17]. Recently, WCR has been reported to evolve resistance to the newest management technology, RNA interference (RNAi) [51]. Future research could aim to leverage automation and robotics technology to establish the high-throughput system for WCR that would accelerate discovery efforts related to novel insecticide compounds and their related products.

Author Contributions: M.P.H. contributed to writing the first draft of the manuscript. M.P.H., M.G.V., T.A.C. and B.E.H. conceptualized the study. M.P.H., M.G.V. and B.E.H. developed the methodology. M.P.H., A.E.P. and R.W.G. performed the experiments. M.P.H. collected data, performed the analyses, and provided the visualization. M.G.V., K.S.S. and B.E.H. provided materials required for the experiments. All authors have read and agreed to the published version of the manuscript.

Funding: Research for this study, in part, was funded by the Small Business Innovation Research (SBIR) Program at the National Institute for Food and Agriculture (NIFA) (grant number 2020-33610-31703).

Institutional Review Board Statement: Not applicable.

Informed Consent Statement: Not applicable.

Data Availability Statement: Not applicable.

Acknowledgments: The authors thank Julie Barry, Amanda Ernwall, Michelle Gregory at USDA-ARS (Columbia, Missouri), and David Tague at University of Missouri for technical assistance. Mention of trade names or commercial products is solely for the purpose of providing specific information and does not imply recommendation or endorsement by the USDA. USDA is an equal opportunity provider and employer. This article reports the results of research only. Mention of a proprietary product does not constitute an endorsement or recommendation for its use by the USDA or the University of Missouri.

Conflicts of Interest: Michael Vella is an employee of Frontier Scientific Services, Newark, Delaware. All other authors declare no competing interests.

References

1. Veres, A.; Wyckhuys, K.A.; Kiss, J.; Tóth, F.; Burgio, G.; Pons, X.; Avilla, C.; Vidal, S.; Razinger, J.; Bazok, R. An update of the Worldwide Integrated Assessment (WIA) on systemic pesticides. Part 4: Alternatives in major cropping systems. *Environ. Sci. Pollut. Res.* **2020**, *27*, 29867–29899. [CrossRef]
2. Wechsler, S.; Smith, D. Has resistance taken root in US corn fields? Demand for insect control. *Am. J. Agr. Econ.* **2018**, *100*, 1136–1150. [CrossRef]
3. Branson, T.F.; Krysan, J.L. Feeding and oviposition behavior and life cycle strategies of *Diabrotica*: An evolutionary view with implications for pest management. *Environ. Entomol.* **1981**, *10*, 826–831. [CrossRef]
4. Moeser, J.; Vidal, S. Nutritional resources used by the invasive maize pest *Diabrotica virgifera virgifera* in its new south-east-European distribution range. *Entomol. Exp. Appl.* **2005**, *114*, 55–63. [CrossRef]
5. Culy, M.D.; Edwards, C.R.; Cornelius, J.R. Effect of silk feeding by western corn rootworm (Coleoptera: Chrysomelidae) on yield and quality of inbred corn in seed corn production fields. *J. Econ. Entomol.* **1992**, *85*, 2440–2446. [CrossRef]

6. Ball, H.J.; Weekman, G.T. Differential resistance of corn rootworms to insecticides in Nebraska and adjoining States. *J. Econ. Entomol.* **1963**, *56*, 553–555. [CrossRef]

7. Meinke, L.J.; Siegfried, B.D.; Wright, R.J.; Chandler, L.D. Adult susceptibility of Nebraska western corn rootworm (Coleoptera: Chrysomelidae) populations to selected insecticides. *J. Econ. Entomol.* **1998**, *91*, 594–600. [CrossRef]

8. Pereira, A.E.; Wang, H.; Zukoff, S.N.; Meinke, L.J.; French, B.W.; Siegfried, B.D. Evidence of field-evolved resistance to Bifenthrin in western corn rootworm (*Diabrotica virgifera virgifera* LeConte) populations in Western Nebraska and Kansas. *PLoS ONE* **2015**, *10*, e0142299. [CrossRef] [PubMed]

9. Pereira, A.E.; Souza, D.; Zukoff, S.N.; Meinke, L.J.; Siegfried, B.D. Cross-resistance and synergism bioassays suggest multiple mechanisms of pyrethroid resistance in western corn rootworm populations. *PLoS ONE* **2017**, *12*, e0179311. [CrossRef]

10. Gassmann, A.J.; Petzold-Maxwell, J.L.; Keweshan, R.S.; Dunbar, M.W. Field-evolved resistance to Bt maize by western corn rootworm. *PLoS ONE* **2011**, *6*, e22629. [CrossRef]

11. Gassmann, A.J.; Shrestha, R.B.; Jakka, S.R.; Dunbar, M.W.; Clifton, E.H.; Paolino, A.R.; Ingber, D.A.; French, B.W.; Masloski, K.E.; Dounda, J.W. Evidence of resistance to Cry34/35Ab1 corn by western corn rootworm (Coleoptera: Chrysomelidae): Root injury in the field and larval survival in plant-based bioassays. *J. Econ. Entomol.* **2016**, *109*, 1872–1880. [CrossRef]

12. Zukoff, S.N.; Zukoff, A.L.; Geisert, R.W.; Hibbard, B.E. Western corn rootworm (Coleoptera: Chrysomelidae) larval movement in eCry3.1Ab+mCry3A seed blend scenarios. *J. Econ. Entomol.* **2016**, *109*, 1834–1845. [CrossRef]

13. Ludwick, D.C.; Meihls, L.N.; Ostlie, K.R.; Potter, B.D.; French, L.; Hibbard, B.E. Minnesota field population of western corn rootworm (Coleoptera: Chrysomelidae) shows incomplete resistance to Cry34Ab1/Cry35Ab1 and Cry3Bb1. *J. Appl. Entomol.* **2017**, *141*, 28–40. [CrossRef]

14. Calles-Torrez, V.; Knodel, J.J.; Boetel, M.A.; French, B.W.; Fuller, B.W.; Ransom, J.K. Field-evolved resistance of northern and western corn rootworm (Coleoptera: Chrysomelidae) populations to corn hybrids expressing single and pyramided Cry3Bb1 and Cry34/35Ab1 Bt proteins in North Dakota. *J. Econ. Entomol.* **2019**, *112*, 1875–1886. [CrossRef]

15. Gassmann, A.J.; Shrestha, R.B.; Kropf, A.L.; St Clair, C.R.; Brenizer, B.D. Field-evolved resistance by western corn rootworm to Cry34/35Ab1 and other *Bacillus thuringiensis* traits in transgenic maize. *Pest Manag. Sci.* **2020**, *76*, 268–276. [CrossRef]

16. Levine, E.; Spencer, J.L.; Isard, S.A.; Onstad, D.W.; Gray, M.E. Adaptation of the western corn rootworm to crop rotation: Evolution of a new strain in response to a management practice. *Am. Entomol.* **2002**, *48*, 94–117. [CrossRef]

17. Gray, M.E.; Sappington, T.W.; Miller, N.J.; Moeser, J.; Bohn, M.O. Adaptation and invasiveness of western corn rootworm: Intensifying research on a worsening pest. *Ann. Rev. Entomol.* **2009**, *54*, 303–321. [CrossRef]

18. EPA. *Memo from EPA on Corn Rootworm Resistance Management*; U.S. Environmental Protection Agency (EPA): Washington, DC, USA, 2016.

19. Meihls, L.N.; Higdon, M.L.; Siegfried, B.D.; Miller, N.J.; Sappington, T.W.; Ellersieck, M.R.; Spencer, T.A.; Hibbard, B.E. Increased survival of western corn rootworm on transgenic corn within three generations of on-plant greenhouse selection. *Proc. Natl. Acad. Sci. USA* **2008**, *105*, 19177–19182. [CrossRef] [PubMed]

20. Nowatzki, T.M.; Lefko, S.A.; Binning, R.R.; Thompson, S.D.; Spencer, T.; Siegfried, B. Validation of a novel resistance monitoring technique for corn rootworm (Coleoptera: Chrysomelidae) and event DAS-59122-7 maize. *J. Appl. Entomol.* **2008**, *132*, 177–188. [CrossRef]

21. Zukoff, S.N.; Ostlie, K.R.; Potter, B.; Meihls, L.N.; Zukoff, A.L.; French, L.; Ellersieck, M.R.; Wade French, B.; Hibbard, B.E. Multiple assays indicate varying levels of cross resistance in Cry3Bb1-selected field populations of the western corn rootworm to mCry3A, eCry3.1Ab, and Cry34/35Ab1. *J. Econ. Entomol.* **2016**, *109*, 1387–1398. [CrossRef] [PubMed]

22. Marrone, P.G.; Ferri, F.D.; Mosley, T.R.; Meinke, L.J. Improvements in laboratory rearing of the southern corn rootworm, *Diabrotica undecimpuncta howardi* Barber (Coleoptera: Chrysomelidae), on an artificial diet and corn. *J. Econ. Entomol.* **1985**, *78*, 290–293. [CrossRef]

23. Sutter, G.R.; Krysan, J.L.; Guss, P.L. Rearing the southern corn rootworm on artificial diet. *J. Econ. Entomol.* **1971**, *64*, 65–67. [CrossRef]

24. Pleau, M.J.; Huesing, J.E.; Head, G.P.; Feir, D.J. Development of an artificial diet for the western corn rootworm. *Entomol. Exp. Appl.* **2002**, *105*, 1–11. [CrossRef]

25. Huynh, M.P.; Meihls, L.N.; Hibbard, B.E.; Lapointe, S.L.; Niedz, R.P.; Ludwick, D.C.; Coudron, T.A. Diet improvement for western corn rootworm (Coleoptera: Chrysomelidae) larvae. *PLoS ONE* **2017**, *12*, e0187997. [CrossRef]

26. Ludwick, D.C.; Meihls, L.N.; Huynh, M.P.; Pereira, A.E.; French, B.W.; Coudron, T.A.; Hibbard, B.E. A new artificial diet for western corn rootworm larvae is compatible with and detects resistance to all current Bt toxins. *Sci. Rep.* **2018**, *8*, 5379. [CrossRef] [PubMed]

27. Huynh, M.P.; Hibbard, B.E.; Vella, M.; Lapointe, S.L.; Niedz, R.P.; Shelby, K.S.; Coudron, T.A. Development of an improved and accessible diet for western corn rootworm larvae using response surface modeling. *Sci. Rep.* **2019**, *9*, 16009. [CrossRef] [PubMed]

28. Meihls, L.N.; Huynh, M.P.; Ludwick, D.C.; Coudron, T.A.; French, B.W.; Shelby, K.S.; Hitchon, A.J.; Schaafsma, A.W.; Pereira, A.E.; Hibbard, B.E. Comparison of six artificial diets for support of western corn rootworm bioassays and rearing. *J. Econ. Entomol.* **2018**, *111*, 2727–2733. [CrossRef] [PubMed]

29. Paramasiva, I.; Sharma, H.C.; Krishnayya, P.V. Antibiotics influence the toxicity of the delta endotoxins of *Bacillus thuringiensis* towards the cotton bollworm, *Helicoverpa armigera*. *BMC Microbiol.* **2014**, *14*, 200. [CrossRef]

30. Visweshwar, R.; Sharma, H.; Akbar, S.; Sreeramulu, K. Elimination of gut microbes with antibiotics confers resistance to *Bacillus thuringiensis* toxin proteins in *Helicoverpa armigera* (Hubner). *Appl. Biochem. Biotechnol.* **2015**, *177*, 1621–1637. [CrossRef] [PubMed]

31. Chu, C.-C.; Spencer, J.L.; Curzi, M.J.; Zavala, J.A.; Seufferheld, M.J. Gut bacteria facilitate adaptation to crop rotation in the western corn rootworm. *Proc. Natl. Acad. Sci. USA* **2013**, *110*, 11917–11922. [CrossRef]

32. Paddock, K.J.; Pereira, A.E.; Finke, D.L.; Ericsson, A.C.; Hibbard, B.E.; Shelby, K.S. Host resistance to *Bacillus thuringiensis* is linked to altered bacterial community within a specialist insect herbivore. *Mol. Ecol.* **2021**, in press. [CrossRef] [PubMed]

33. Huynh, M.P.; Hibbard, B.E.; Lapointe, S.L.; Niedz, R.P.; French, B.W.; Pereira, A.E.; Finke, D.L.; Shelby, K.S.; Coudron, T.A. Multidimensional approach to formulating a specialized diet for northern corn rootworm larvae. *Sci. Rep.* **2019**, *9*, 3709. [CrossRef] [PubMed]

34. Cohen, A.C. *Insect Diets: Science and Technology*, 2nd ed.; CRC Press, Taylor & Francis Group: Boca Raton, FL, USA, 2015.

35. Damodaran, S.; Parkin, K.L.; Fennema, O.R. *Fennema's Food Chemistry*; CRC Press: Boca Raton, FL, USA, 2007.

36. Huynh, M.P.; Bernklau, E.J.; Coudron, T.A.; Shelby, K.S.; Bjostad, L.B.; Hibbard, B.E. Characterization of corn root factors to improve artificial diet for western corn rootworm (Coleoptera: Chrysomelidae) larvae. *J. Insect Sci.* **2019**, *19*, 20. [CrossRef] [PubMed]

37. O'Connell, K.P.; Kovaleva, E.; Campbell, J.H.; Anderson, P.E.; Brown, S.G.; Davis, D.C.; Valdes, J.J.; Welch, R.W.; Bentley, W.E.; van Beek, N.A. Production of a recombinant antibody fragment in whole insect larvae. *Mol. Biotechnol.* **2007**, *36*, 44–51. [CrossRef]

38. Allen, D.M. Mean square error of prediction as a criterion for selecting variables. *Technometrics* **1971**, *13*, 469–475. [CrossRef]

39. Myers, R.H.; Montgomery, D.C.; Anderson-Cook, C.M. *Response Surface Methodology: Process and Product Optimization Using Designed Experiments*, 4th ed.; John Wiley & Sons: New York, NY, USA, 2016.

40. Anderson, M.J.; Whitcomb, P.J. Using graphical diagnostics to deal with bad data. *Qual. Eng.* **2007**, *19*, 111–118. [CrossRef]

41. Damodaran, S. Amino acids, peptides and proteins. In *Fennema's Food Chemistry*; CRC Press, Taylor & Francis Group: Boca Raton, FL, USA, 2008; Volume 4, pp. 425–439.

42. Vanderzant, E. Improvements in the rearing diet for *Chrysopa carnea* and the amino acid requirements for growth. *J. Econ. Entomol.* **1973**, *66*, 336–338. [CrossRef]

43. Vanderzant, E. An artificial diet for larvae and adults of *Chrysopa carnea*, an insect predator of crop pests. *J. Econ. Entomol.* **1969**, *62*, 256–257. [CrossRef]

44. Griffin, J.; Lindig, O.; McLaughlin, R. Flash sterilizers: Sterilizing artificial diets for insects. *J. Econ. Entomol.* **1974**, *67*, 689. [CrossRef]

45. Oyediran, I.O.; Hibbard, B.E.; Clark, T.L. Prairie grasses as hosts of the western corn rootworm (Coleoptera: Chrysomelidae). *Environ. Entomol.* **2004**, *33*, 740–747. [CrossRef]

46. Siegfried, B.D.; Vaughn, T.T.; Spencer, T. Baseline susceptibility of western corn rootworm (Coleoptera: Crysomelidae) to Cry3Bb1 *Bacillus thuringiensis* toxin. *J. Econ. Entomol.* **2005**, *98*, 1320–1324. [CrossRef] [PubMed]

47. Pereira, A.E.; Huynh, M.P.; Sethi, A.; Miles, A.L.; Wade French, B.; Ellersieck, M.R.; Coudron, T.A.; Shelby, K.S.; Hibbard, B.E. Baseline susceptibility of a laboratory strain of northern corn rootworm, *Diabrotica barberi* (Coleoptera: Chrysomelidae) to *Bacillus thuringiensis* traits in seedling, single plant, and diet-toxicity assays. *J. Econ. Entomol.* **2020**, *113*, 1955–1962. [CrossRef] [PubMed]

48. Pereira, A.E.; Huynh, M.P.; Carlson, A.R.; Haase, A.; Kennedy, R.M.; Shelby, K.S.; Coudron, T.A.; Hibbard, B.E. Assessing the single and combined toxicity of the bioinsecticide Spear® and Cry3Bb1 protein against susceptible and resistant western corn rootworm larvae (Coleoptera: Chrysomelidae). *J. Econ. Entomol.* **2021**, in press. [CrossRef]

49. Tillman, P.G.; McKibben, G.; Malone, S.; Harsh, D. *Form-Fill-Seal Machine for Mass Rearing Noctuid Species*; Mississippi Agricultural and Forestry Experiment Station: Mississippi State, MS, USA, 1997.

50. Paddock, K.J.; Robert, C.A.; Erb, M.; Hibbard, B.E. Western Corn Rootworm, Plant and Microbe Interactions: A Review and Prospects for New Management Tools. *Insects* **2021**, *12*, 171. [CrossRef] [PubMed]

51. Khajuria, C.; Ivashuta, S.; Wiggins, E.; Flagel, L.; Moar, W.; Pleau, M.; Miller, K.; Zhang, Y.; Ramaseshadri, P.; Jiang, C. Development and characterization of the first dsRNA-resistant insect population from western corn rootworm, *Diabrotica virgifera virgifera* LeConte. *PLoS ONE* **2018**, *13*, e01970. [CrossRef]

MDPI
St. Alban-Anlage 66
4052 Basel
Switzerland
Tel. +41 61 683 77 34
Fax +41 61 302 89 18
www.mdpi.com

Insects Editorial Office
E-mail: insects@mdpi.com
www.mdpi.com/journal/insects